EQUALIZED & SYNCHRONIZED PRODUCTION

EQUALIZED & SYNCHRONIZED PRODUCTION

....

The High-Mix Manufacturing System That Moves Beyond JIT

TOSHIKI NARUSE

Contributing Authors
Kenichi Morii
Kunio Shibata
Tsutomu Iwabuchi

PRODUCTIVITY
productivity press

PRODUCTIVITY PRESS • NEW YORK, NY

Additional copies of this book are available from the publisher. Discounts are available for multiple copies through the Sales Department (800-394-6868). Address all other inquiries to:

Productivity Press
444 Park Avenue South, Suite 604
New York, NY 10016
United States of America
Telephone: 212-686-5900
Telefax: 212-686-5411
E-mail: *info@productivityinc.com*

Cover design by Gary Ragaglia
Page composition by William H. Brunson Typography Services
Printed by Malloy Lithographing in the United States of America

Library of Congress Cataloging-in-Publication Data
Naruse, Toshiki.
   Equalized and Synchronized Production: The High-Mix Manufacturing System that moves beyond JIT/ Toshiki Naruse.
        p. cm.
      ISBN 1-56327-252-0
      1. Production management.   2. Manufacturing processes.
   3. Production control.   I. Title.
   TS155.N1178   2003
   658.5—dc21
                                                            2002154462

06 05 04 03   5 4 3 2 1

Contents

Preface

Companies all over the world have studied and adopted the Just-in-Time production system (JIT), as exemplified by the Toyota Production System. But is JIT really the best type of production system for every manufacturing company? At JMAC we began exploring this simple question and found that for end buyers, such as automobile manufacturers that process or assemble the final products for merchants and buyers, JIT is a very useful system indeed. But JIT is not necessarily useful for suppliers that supply parts and materials to these types of buyers. In fact, many of these suppliers who must adhere to the strict requirements of JIT imposed on them by their buyer's JIT program, are being exploited. In response to this, JMAC has developed a new production system for suppliers, the Equalized and Synchronized Production (ESP) System, that revolutionizes production management.

Though the practice of JIT includes various types of procedures and programs for improvement, it essentially achieves the same results as synchronization. While this is indeed very important, many problems are likely to arise when all the manufacturing companies in a supplier chain try to practice JIT strictly. For example, for an automotive parts and materials supplier to comply fully with the automobile manufacturer's JIT program, the supplier must physically locate its plants either within the manufacturer's site or nearby. Then it must supply parts and materials strictly according to the automobile manufacturer's final assembly schedule, so that parts and materials are delivered just in time for the vehicle assembly processes. In some cases, suppliers have built plants so close to their buyers' plants that they can use conveyor belts to deliver parts to the buyer's final assembly line.

However, very few automotive parts suppliers go to such extremes to comply fully with their buyer's JIT production systems. Instead, they often end up building product warehouses close to the buyer so they can truck the products (parts and materials) to the buyer's final assembly processes just in time and in just the order required by the buyer's final assembly schedules.

Also, very few automotive parts suppliers ship to only one factory owned by only one buyer. Usually, suppliers ship their parts and materials to several different automobile manufacturers and perhaps several plants belonging to each one. This means that suppliers are faced with the daunting task of coordinating and delivering their shipments according to the JIT production needs of multiple production facilities belonging to several different end buyers.

Given this situation, it is almost impossible for suppliers to implement synchronization well enough to meet the production requirements (specific volumes and delivery deadlines for specific product items) of each production facility owned by each end buyer they serve. As a result, suppliers must invest great sums in synchronizing their operations with those of their buyers and this investment burden can be so great as to threaten the survival of the supplier.

The ESP Production System was born in 1985, as a result of a consulting project for an automotive electrical components manufacturer that was serving several automobile manufacturers at their respective production facilities. Each of these buyers required this supplier to meet its particular JIT production timetable.

By 1985, this automotive electrical components manufacturer had already built and was operating its own JIT program in an effort to accomplish the difficult task of meeting various buyers' JIT production needs. This arrangement. however, resulted in many production management problems for this supplier, such as an abundance of loss and abnormalities that occurred in their production processes.

Through this consulting project, I and the other consultants at JMAC became convinced that the best type of production system for suppliers was one that emphasized the concept of synchronization, and incorporated the concept of equalization. In fact, it was through this consulting project that we proved we could incorporate both of these concepts without having to dismantle the existing JIT production and build from scratch. Instead, we built ESP on top of JIT. In other words, you can adopt a production system based on the concepts of synchronization and equalization for production scheduling functions and the underlying purchasing functions, as well as use this same approach to incorporate the main work of making improvements needed to conduct production management tasks and factory-floor operations. In short, the ESP Production System takes synchronization as its foundation and then adds equalization. It's as simple and obvious as that.

Later, JMAC consultants further refined the ESP program as they helped other suppliers, chiefly automotive parts suppliers, develop their own ESP Production Systems. Eventually this grand experiment was widened, introducing ESP at companies in nonautomotive industries. The upshot is that we at JMAC became convinced that the ESP approach is the best approach for suppliers in various industries, as well as for companies that distribute their products from warehouses via sales companies and agencies, such as general machinery parts and precision machinery parts. ESP can also be the best approach for end buyers that produce food products, medical supplies, and so on.

Because of remarkable progress in information technology (IT) in recent years, there has been a large increase in the number of companies that are implementing Enterprise Resource Planning (ERP) and Supply Chain Management (SCM). The goal for these companies is to realize an efficient and quick management system that is consistent and transparent to the global market. In other words, these companies are required to respond to worldwide competition (mega-competition) and speedy management (as the life-cycle of various kinds of industrial products is getting shorter), as well as to create valuable and differentiated products. It is getting more and more difficult to differentiate products only by product function, price and quality. Supplier companies must customize and provide products in a more timely manner by shortening delivery speed. They are not only required to deliver products Just-in-Time, but they have to achieve the "advanced" Just-in-Time (more variety, less quantity, and less time). In order to achieve this, they have to free themselves from informal methods that depend on people and re-establish a management method and operation process that works systematically. It is necessary to analyze and recreate one's management model in order to change the structure, but, in many cases, companies use "standard process" that is provided in ERP and SCM.

Many of these companies have not achieved the anticipated results. This may have been due to their failure to fully consider how to align their new ERP or SCM system with their company's own particular set of functions and operations. For example, the production management field uses ERP and SCM systems based on Materials Resource Planning (MRP) methods that have remained almost unchanged for 30 or 40 years. This may be one of the factors that keeps various manufacturing companies from being able to reform the methods of the production management and production divisions.

Even before ERP and SCM gained popularity, the information technology and production management divisions at companies that introduced MRP stuck to a certain set of rules, rules registered as part of the MRP approach. The main reason for this might have been that it was too much of a hassle to change the set of rules. Companies that develop or provide MRP systems will emphasize the need to change and reregister the rules whenever it is deemed appropriate, and will even go as far as to say that such flexibility is a fundamental part of MRP, a function that should be used to the utmost. But the fact remains that very few companies go to the trouble of putting this flexibility into practice. As a result, MRP-based operations have been characterized as a rigid set of production scheduling and purchasing functions.

When this occurs, other negative factors tend to appear, including a weakening of production management's recognition for the need for factory-floor improvements, such as product model changeover improvements, and their desire to make such improvements. It goes without saying that a manufacturing company has a critical obligation to constantly strive for innovative development of distinctive technologies, including production methods and management techniques, to reduce costs, improve cash flow, and boost profits. However, as explained above, when production management adopts MRP-based ERP and SCM systems, they work to constrain such technological innovation, especially in production and management technologies. Or they at least make front-line operations more complicated and difficult.

ESP has strength precisely in the area where ERP and SCM (i.e., MRP for production management) are weakest. That is, it complements each company's own production management functions, particularly production scheduling and purchasing functions, in the way that best suits each company. In addition, the ESP approach also complements ERP and SCM by acting as a powerful force to encourage innovations in engineering and manufacturing. It is no exaggeration to say that by adopting ESP, a company can round out its advanced ERP and/or SCM system so that it incorporates a full-fledged set of the system's functions. Not only that, your company will cease to be subservient to the will of its buyers and, instead, become a powerful new organization operating according to its own production system.

It's the authors hope that this book will help companies recognize the value of the ESP Production System and assist them in implementing this approach. I'm confident that ESP will help your company.

Acknowledgments

Finally, the authors would like to express their deep gratitude to Ms. Maura May and Ms. Miho Matsubara of Productivity Press for their unflagging efforts and valuable advice concerning this book. They would also like to take this opportunity to sincerely thank Mr. Kazuya Uchiyama, who assisted in the production of the Japanese manuscript and who served wonderfully as a liaison between Productivity Press and JMAC. We also thank Gary Peurasaari for the content editing and the work of Ms. Miho Matsubara and Michael Sinocchi in shaping the manuscript.

November, 2002
Toshiki Naruse
Principal Consultant
Japan Management Association Consultants

The Birth of a New Production System—A Case Study

The principles behind the ESP Production System evolved from a few concepts that JMAC developed during a consulting project in 1985 for an automotive electrical components manufacturer. This company was supplying components to nearly every automaker in Japan. Consequently, the company was receiving orders from each automaker's car assembly plants located in various parts of Japan. Deliveries were assigned not only to certain buyers, but also to certain production lines at certain plants. To help accommodate this, this supplier took the initiative early on to thoroughly research and correctly adopt the Toyota Production System, or what is commonly referred to as Just-in-Time (JIT).

As a result of the production process improvements, this supplier installed several assembly lines for particular automakers and car models and then set up assembly lines corresponding to the automobile manufacturers and production lines from which orders had been received. Now the supplier could deliver products from each line straight to their respective destinations.

This supplier also used a kanban system to distribute in-house production specifications. They built and operated their kanban in accordance with the basic principles of kanban systems. However, the supplier's production processes were plagued by an unstable rate of production efficiency, which varied in tandem with the daily changes in production output. The news varied between good and bad with each passing day, with managers making daily pronouncements such as "Congratulations. Everyone worked hard today. Our operator efficiency rating was 90 percent," or "Looks like we failed to hit the right pace. Our operator efficiency rating was only 70 percent, which means we'll have to put in two hours of overtime tomorrow." When production output fell short of the amounts needed for deliveries, unexpected needs for overtime work arose, and when production reached the needed amounts in less than a full day, standby time (= waste) occurred.

Even though the supplier could handle several dozen types of products (item numbers) on each of its assembly lines, its manufacturing costs were on the rise for a number of reasons, such as:

- Dealing with the production sequence of item numbers varying from day to day.
- Dealing with the number of production specifications per item number varying each time as production specifications were issued four times a day. As a result, in order to produce small lots of finished products, the company divided each working day into four two-hour segments of production and issued production specifications four times (cycles) each day.
- Having to spend a lot of labor hours on setting up parts because of the above variations. For example, parts had to be counted and arranged for supply to their respective assembly lines in the exact amounts needed for each item number in order to avoid assembly errors such as assembly using the wrong parts.

The supplier had already done everything it could to improve its direct operations and now they were faced with the need to improve their semidirect operations in order to meet their cost-cutting targets to remain competitive. So, early in the summer of 1985 the supplier contacted JMAC to solve a problem: they were spending way too many labor hours in semidirect operations for final assembly processes (operations such as supplying parts or managing lines). And they sorely needed to cut back on semidirect operations-related labor in order to cut costs. So they asked for our help in determining how to go about improving their semidirect operations while reducing their labor needs for such operations. So the original objective of our assignment was to increase the efficiency of the supplier's assembly lines.

TWO ORIGINAL CONCEPTS BEHIND ESP

Before discussing the specifics of this consulting project, we will first look at the two original concepts behind the ESP Production

System that we devised during this 1985 consulting project (see Figure 1-1). The basic idea is that a supplier cannot succeed in fulfilling the mission of its production division (see Figure 1-2), which is to produce according to the buyer's needs, as long as production (each production line) is not dedicated completely to a particular end buyer.

Original Concept One	Production geared strictly to customer needs is inefficient.
Original Concept Two	To fulfill the production division's mission, daily production output and production sequences must be stabilized, with production output equalized among the various item numbers.

Figure 1-1. Two Original Concepts behind the ESP Production System

The Production Division's mission is: • To maximize production efficiency by making and maintaining improvements toward that end. • To minimize inventory by working toward the goal of zero inventory.

Figure 1-2. Production Division's Mission

- **Original concept one.** This concept refers to a situation in which a supplier tries to simply turn out products as needed to meet the buyer's orders, which means that any fluctuation (peaks and valleys) in the amount of items ordered will be directly reflected in the production processes, making it impossible for the supplier to control its production efficiency. In other words, if the supplier is going to be responsible for managing its own production efficiency, it cannot afford to link its production schedule directly to its orders from buyers. Instead, it must make use of product inventory to separate its production schedule from such a direct link to the buyer's orders.
- **Original concept two.** This concept refers to the need for suppliers to reconfirm the production division's mission and to ask what kinds of methods it must use to fulfill this mission. Generally, the manufacturing division's resources (personnel, equipment, etc.)

are fixed for a certain period of time. Under these conditions, the daily production output (or, to be precise, the production load's labor hours) must be stabilized in order to fill the orders sent to the Manufacturing Division. In addition, in order to make production more efficient (with assured volume and stable quality) and to make the production setup and clean-up operations more efficient, the supplier must set a stable production sequence for the various types of products (item numbers), with equalized production volume for each item number.

These two original concepts were the starting point for JMAC's efforts to liberate the supplier's production scheduling from any direct connection to the inflow of received orders. They accomplished this by establishing a limited amount of product inventory (limited inventory) and a production schedule in which the production sequence had 1) products (item numbers) arranged so as to maximize production efficiency, and 2) equalized production output specified for each item number. During this consulting project, the methods that we used to enable the establishment of this kind of production schedule, such as setting production planning standards, creating a production schedule planning process, and improving production processes, became the foundation (principles and guarantees) upon which we later built the ESP Production System. These principles and guarantees will be discussed further in Chapter 2.

THE SUPPLIER'S PREDICAMENT—A CASE STUDY

When we first encountered this electronic components supplier, it had devised 22 final assembly lines to service each auto manufacturer and specific car models, with products being delivered to auto plants and assembly lines all over Japan. In addition to these final assembly processes, the supplier operated several upstream processes in house, such as molding, parts machining, and component assembly processes, and also had several outsourced processes. (See Figure 1-3.)

This supplier employed some 200 people at the final assembly lines, and 36 of these workers were involved in semidirect operations. The semidirect operations were mainly focused on the following six areas:

1. Gathering parts for each line.
2. Transporting and distributing parts to each line.
3. Transporting products to each line (transporting to the shipping area).

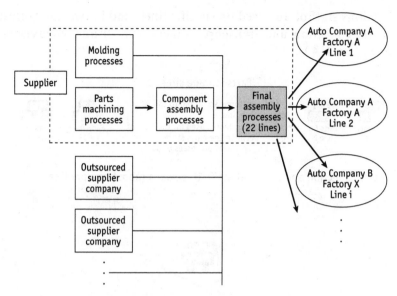

Figure 1-3. Overview of Production Processes at the Supplier

4. Handling the processing of parts kanban (used when ordering items from in-house upstream processes or from outsourced suppliers).
5. Handling the processing of product kanban (the final specifications for the assembly schedule and shipping schedule).
6. Performing supervisory tasks for each line, including efforts to make improvements.

The system for supplying goods to these final assembly processes included 22 separate parts storage sites—one for each of the 22 final assembly processes—as well as a kanban system that was used by upstream processes and outsourced suppliers to request parts from these storage sites. Each parts storage site maintained an inventory level that exactly matched the number of parts required by the final assembly lines. The parts were put in locations marked for specific final assembly lines and then were delivered when needed by those lines.

Products turned out by final assembly lines were temporarily stored in pass boxes according to the instructions received from the buyers (the auto manufacturers), and were then conveyed to the shipping area (product storage site). (The above is a simple summary of how goods flowed to and from the final assembly processes.) Specific lines were always the units used to manage the final assembly processes. Accordingly, the supplier also assigned its semidirect operations staff of 36 based on specific lines, and in two categories: 1) parts collection and transport workers, and 2) line supervisors. (See Figure 1-4.)

Parts storage sites

Parts for line 1   Parts for line 2

- Parts are stored according to the destination line.
- A kanban system is used for the flow of goods between the parts storage sites and the upstream processes or parts suppliers.

Parts for line 21   Parts for line 22

Parts collection and transport staff: 18 persons (about one per line).

Assembly processes (for all 22 lines)

Line 1 | Line 2 | Line 3 | Line 4 | Line 5 | Line 6 | Line 7 | Line 8 | Line 9 | Line 10 | Line 11 | Line 12 | Line 13 | Line 14 | Line 15 | Line 16 | Line 17 | Line 18 | Line 19 | Line 20 | Line 21 | Line 22

Line supervisors: 18 persons (about one per line)

Shipping area (product storage site)

- Products are stored according to their destination lines.
- Products are stored according to how they are packed for delivery (using pass boxes designated for particular final companies or destination lines, with a designated amount of products in each pass box).
- A kanban system is used with the final companies to manage deliveries from product storage sites to final companies.

The basic approach to process management is one of "line-specific management."
*There are 36 persons employed in indirect operations (semidirect operations). These include 18 line supervisors; 18 parts collection and transportation staff.

Figure 1-4. Overview of Assembly Processes and Assignment of Semidirect Operations Staff

Implementing JIT Failed to Achieve Stable Production Efficiency

As a result of implementing the Toyota Production System, the supplier thought they had improved their final assembly process with its 22 assembly lines each servicing one auto manufacturer or car model. Now orders from each buyer's assembly plant could be sent to the supplier's corresponding assembly line for production and delivery. The supplier now employed a kanban system that used withdrawal kanban and outside supplier kanban to manage the procurement of parts from in-house and outside sources. In addition, they used other kanban (product kanban) to manage the issuing of production specifications to final assembly processes and the delivery of products to the various automobile manufacturers. When we arrived, they were correctly running the various kanban systems, yet, at the final assembly processes, they were still unable to maintain stable production efficiency due to the constant, daily fluctuation in production volume. In fact, as mentioned above, it was not unusual for productivity to change sharply in either direction from day to day—from 90 percent of operator productivity to 70 percent the next day. Whenever their production yield was not enough to meet an upcoming delivery deadline, they were forced to perform unplanned overtime to make up the shortfall. The last-minute changes in work hours to make up for production shortfalls tended to result in more operator standby waste.

Given that each production line handled dozens of product models (item numbers), factors such as daily changes in the sequence in which certain item numbers were produced, or changes in the number of units indicated for certain item numbers in the production specifications (issued four times daily), meant that 36 employees needed many labor hours to manage the collection and transport of parts. This work included confirming the quantities of each set of parts to be sent for assembly per item number, so as to avoid defects due to misassembled components or components that contained the wrong parts. Little wonder that this supplier was vexed by rising manufacturing costs and needed help. We will now discuss four phases of assessing the problem and developing a way to switch from JIT to a synchronized and equalized production system.

PHASE ONE: ASSESSING THE CURRENT SITUATION FOR SEMIDIRECT OPERATIONS AT THE FINAL ASSEMBLY PROCESSES

In response to the supplier's request for improving the semidirect operations (parts supply and line management operations) at the final assembly processes and reducing labor requirements for semidirect operations, JMAC began its assessment of the situation by researching and analyzing in detail the current operations and work methods. To understand, in general, and then later, in detail, the semidirect operations and the methods used by the supplier, we paired each semidirect operations worker with a researcher who conducted a continuous operation analysis. To assist the improvement efforts of each semidirect operations worker being studied, these continuous operation analyses included a clarification of each worker's work objectives and the behavior (work) that each worker performed to achieve those objectives. We did this because analyzing the work of each semidirect operations worker (collect parts, count parts, transport parts, walk, set up parts, etc.) without first understanding its purpose made it unlikely we could devise improvements.

Each action a worker took in his or her semidirect operations—whether collecting parts, counting parts, transporting parts, walking, setting up parts, or whatever—had to have a purpose. In other words, workers are not likely to be doing anything for no reason at all. This is one aspect in which semidirect operations differ from direct operations. Direct operations already have a clear purpose. Therefore, we needed only to observe the actions being performed in those operations before working to devise improvements.

By contrast, we couldn't devise improvements or reduce labor requirements for semidirect operations unless we first identified their purpose. Therefore, we carried out preliminary observations and drafted a list of work units involved in semidirect operations. Next, we determined how the work units of semidirect operations fit into one or more of 18 work-objective categories, which yielded a total of 297 work (action) items based on the 18 work objectives (see Figure 1-5 for an example of that study).

In Figure 1-7 you will see charts illustrating the results of analyses that were performed to determine the work units involved in semidirect operations, as well as the methods and proportional time requirements. The first thing we understood from these results was

Work Unit Configuation Chart			
I Parts supply operation	II 1. Supply of parts to line-specific storage shelves or lines	III 1. Supply parts from line-specific parts stroage sites to line-specific storage shelves	IV -1 Walk to line-spedific parts storage site -2 Pick up parts container -3 Carry parts container -4 Place parts container on line-specific storage shelf -5 Remove lid from parts container -6 Carry lid from parts container -7 Put down lid from parts container
		2. Supply parts from cart to line-specific storage shelves	-1 Walk to cart -2 Pick up parts container -3 Carry parts container -4 Place parts container on line-specific storage shelf -5 Remove lid from parts container -6 Carry lid from parts container -7 Put down lid from parts container
		3. Count parts and supply parts to line-specific storage shelves or lines	-1 Count parts -2 Wrap with rubber band -3 Carry parts or parts box -4 Transport parts or parts box on cart
		4. Supply parts in plastic bag to line-specific storage shelves 5. Supply parts in plastic bag to line 6. Arrange parts containers on line-specific storage shelves 7. Sort and rearrange parts 8. Make arrangements 9. Standby	
	2. Transport parts container	1. Check parts configuration list	-1 Walk to cart or shelf -2 Find and pick up parts configuration list -3 Check parts configuration list -4 Put down parts configuration list -5 Make arrangements -6 Standby
		2. Move empty cart from line to parts storage site or other specified site	-1 Load parts container onto cart -2 Walk to empty cart -3 Move empty cart -4 Walk away from empty cart -5 Make arrangements -6 Standby
		3. Carry parts container from parts storage site to cart	-1 Walk to parts storage site -2 Check loaded parts and item numbers on parts -3 Pick up parts container -4 Carry parts container -5 Load parts container onto cart -6 Leave parts container on cart and walk (to front of cart) etc.

Figure 1-5. Excerpt From Configuration Chart of Work Units in Semidirect Operations

that the work units in semidirect operations can be generally categorized into the following four purposes (see Figure 1-6).

> 1. Actions performed to supply parts to the line.
> 2. Actions performed to transport something.
> 4. Actions performed to collect parts.
> 5. Actions performed by line supervisor.

Figure 1-6. Purposes of Actions Performed by Semidirect Operations Worker

The second thing we understood was that fully 56 percent of the actions performed by semidirect operations workers had some kind of inherent loss (see Figure 1-7). Method-related loss accounted for only 22 percent of the 56 percent, while the other 34 percent was loss inherent in the production system, which has nothing to do with the method used by workers.

We were not at all quick to reach the conclusion that so much loss was inherent in the supplier's production system. In fact, we came to this conclusion only after reviewing the results of our analysis of the series of actions performed by semidirect operations workers, and after applying the E-C-R-S principles of improvement (Eliminate, Combine, Rearrange, Simplify) to each action as a means of devising improvements. In other words, we came to this conclusion after a thorough analysis of semidirect operations.

As a thought experiment during this process of analysis and research, we asked ourselves whether the supplier's current production system might be based on the assumption that we could not make fundamental improvements. This helped us to recognize that much of the semidirect operations currently being performed consisted of irregularly repeated counting, arranging, and carrying work whose purpose is simply to supply just the right amount of parts in just the order required for assembly. In addition, we were able to confirm that these irregularly repeated actions were rooted in the following characteristics of the assembly line.

- No one knows the specifics, such as when, what, and how many products to assemble, until just before assembly begins.
- The quantity of items to be assembled varies widely even among the same product item numbers.

This led us to reexamine how much loss could be avoided by synchronizing and equalizing production. We figured that as much as 34 percent of the semidirect operations work could be

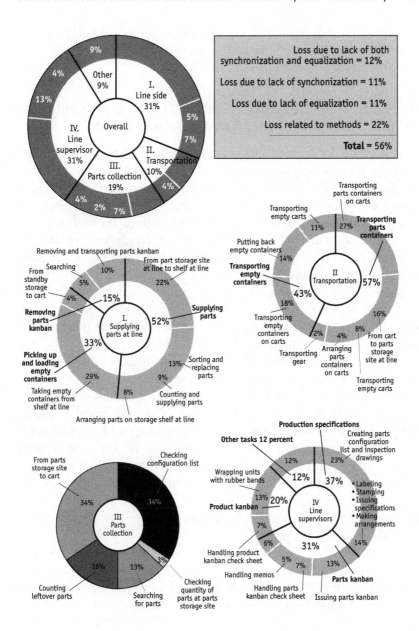

Loss due to lack of both
synchronization and equalization = 12%

Loss due to lack of synchonization = 11%

Loss due to lack of equalization = 11%

Loss related to methods = 22%

Total = 56%

Figure 1-7. Work Units in Semidirect Operations and Proportional Time
Requirements (Overall and by Work Objectives)

identified as loss that occurs because the supplier's production had
not been synchronized and equalized. The breakdown of this 34
percent is as follows.

- 12 percent loss due to lack of both synchronization and equalization.
- 11 percent loss due to lack of synchronization.
- 11 percent loss due to lack of equalization.

PHASE TWO: DEALING WITH THE DIFFICULTIES IN SWITCHING FROM JIT TO A SYNCHRONIZED AND EQUALIZED PRODUCTION SYSTEM

Now that we determined that we could gain a 34 percent improvement by introducing a system of synchronized and equalized production, if we also made work-methods improvements we stood to gain a whopping 56 percent improvement, thereby eliminating loss. Making these improvements, however, would require a major change in thinking. After all, many people firmly believe that JIT (with its kanban system) is the best production system around, and many companies have been doing their utmost to implement JIT with simple and honest adherence to its basic principles. At companies such as these, there is bound to be some stiff resistance to the idea of giving up the kanban system in favor of some new production system, no matter what kinds of benefits are promised.

Figure 1-8 lists six key points of the synchronized and equalized production system proposed for this supplier, and compares and contrasts these points with characteristics of the kanban system.

It should be stated that the purpose of the proposed improvements was not to debate the relative advantages and disadvantages of different production systems. In the case of our improvement themes (our consulting issues) for this particular supplier, we had no intention to hold an abstract debate about production systems, but rather were seeking solutions to specific problems that were occurring at the company. We simply focused on these comparison points as part of our investigation.

As part of this effort, we designed operations that made use of a synchronized and equalized production system and listed the benefits to be gained by adopting this production system so as to help our clients recognize its superiority.

PHASE THREE: MAKING IMPROVEMENTS IN SEMIDIRECT OPERATIONS

Figure 1-9 lists the effects of synchronized and equalized production systems and basic points of adopting such a system by the supplier. Figure 1-10 summarizes the ways in which semidirect

No.	Key points of synchronized and equalized production system	Similarities/differences vs. kanban system
1	Input sequence of product item numbers is determined for each line.	Input sequence is set in advance as part of quality assurance; no significant differences.
2	Production volume per production run of each product item number in each line is equalized (volume is fixed at a certain time).	Production volume per production run of each product item number should always be based on the number of withdrawn kanban (shipped volume); significant difference exists.
3	A basic master plan (ESP production pattern) is created based on input sequences and equalization.	Production schedule should be proposed based on withdrawn kanban (= the amount of confirmed orders received); significant difference exists.
4	Incorporate received buyer orders into the basic master plan (ESP production pattern), create a plan for a regular workday's maximum operation rate (includes a zero missing items plan, an overproduction prevention plan, and a load feed-forward plan), and issue production specifications to the assembly processes.	
5	Apply the principle of limited inventory to separate buyers from the production schedule for the final assembly processes.	Product inventories (safe inventory levels) are still being maintained. There are big differences in the concepts behind the two inventory systems, but there are only small differences in the way the inventory systems are operated.
6	The Planning Department is wholly responsible for establishing and operating the production schedule, with semidirect operations workers acting as agents in implementing the production schedule. Semidirect operations workers switch from a division of labor based on specific assembly lines to one based on specific functions.	Up to now, semidirect operations workers have been responsible for final adjustments and other discretionary matters, and they feel uneasy about surrendering that responsibility to the Planning Department.

Figure 1-8. Six Key Points of Synchronized and Equalized Production System

operations are greatly improved and compares the synchronized and equalized production system with the supplier's current system. By implementing the improvements shown in Figure 1-10 the supplier would be able to reduce the number of semidirect operations workers from the current 36 workers to just 22 workers.

---

1. The main theme in the current improvement project is to use the synchronized and equalized production system as a tool for substantial simplification of quantity management as part of semidirect operations at final assembly processes, and as a means of establishing more rational production management that enables current production conditions to be understood at a glance.

2. This synchronized and equalized production system is offered as a solution to some of the contradictions inherent in the supplier's current kanban system, i.e., as a superior alternative production system for suppliers.

3. While the supplier's kanban systems are *parts order receiving systems* that are intended to make ordering and delivery of parts run more smoothly, such kanban systems can also be thought of as synchronized systems. Obviously, this idea of synchronization is one of the chief features of a synchronized and equalized production system.

4. Another chief characteristic of a synchronized and equalized production system is that it does away with management of parts quantities at intermediate processes, both in-house and at outside vendor companies. As a result, it eliminates the supplier's complicated tasks of processing kanban, handling production specifications, and managing quantities of parts. In short, it enables you to manage production with a minimum of management operations.

5. Since this system establishes equalized production of assembled products and the parts they contain, shortages or excesses of intermediary parts do not occur. In addition, it expands the volume control units for parts from the current *volume control of part units* to *volume control of product units*, which makes the volume control aspect of production management much simpler and more reliable.

6. A synchronized and equalized production system strengthens production planning and production organization functions, and this makes the production system more adaptable to production changes.

Figure 1-9. Effects of Synchronized and Equalized Production System and Basic Points for Adopting It

Loss was estimated at 56 percent and the staff reduction to 22 workers meant that the supplier could reduce loss by about 40 percent by a practical improvement plan that left a margin of error. One might add in jest that, given the advanced state of the IT industry

Synchronized and equalized production system

Parts storage sites
III Collection of parts

II Transport

I At side of line

IV Line supervisor

III Collection of parts
- Parts sent from upstream processes or outside suppliers are put in a standby storage site, where they are arranged according to cycle, input sequence, or product item number.
- Parts are moved from the standby storage sites and are loaded onto carts (from right to left in the diagram).

II Transport
- Carts loaded with containers of collected parts are moved near the assembly line.
- The line-side worker in charge of supplying parts moves the cart loaded with parts containers to the position for supplying parts to the assembly line.

I Supply of parts at side of line
- The line-side worker supplies parts to the line from the loaded cart that was delivered by the transport worker.
- As parts containers are emptied, the line-side worker loads the empty containers onto another cart.
- The line-side worker removes kanban.

IV Line supervisor
- The line supervisor is responsible for handling problems with product kanban (such as ordering line stoppages).
- The line supervisor processes the kanban for the next day's products.
- The line supervisor provides support for workers who are absent, late, or leaving early.
- The line supervisor is in charge of line progress management.
- The line supervisor provides guidance and suggests improvements for operations.
- Other

Points

Equalization
1. Equalization of planning (Equalization of production scope = strengthening of planning functions).
2. Equalization of capacity (Amount of product capacity = amount of part capacity).

Synchronization
1. Synchronization of parts received or assembly into products.
2. Synchronization of parts supplied to assembly lines.

Change parts supply operations so that they are *assigned based on functions, with parts supply workers handling multiple lines.*

Parts at parts storage sites will be stored *according to the input sequence and the target products.*

III Collection of parts
- Collection of required quantities at parts storage site
- Handling of leftover (stray) parts
- Searching for parts
- Confirmation of parts configuration table
- Arrangement of parts in FIFO (first-in, first-out) order

IV Line supervisor
- Data entry on product and parts kanban check sheets
- Writing and distribution of memos
- Quantity checking of important parts

Current system

Parts storage sites
Parts are stored:
- Per line
- Per part type
- Per part item number

line   line   line

* Operations are configured with two workers per line, but there are no clear distinctions among *line-side operations, transport operations, parts collection operations, and line management by line supervisors.* Instead, each pair of workers just focuses on supplying parts in order to prevent line stoppages.

Abnormalities

I Line-side operations
- Parts sorting operations
- Parts counting and supply operations
- Confirmation of counts at line-side parts storage sites
- Arrangement of parts containers at line-side parts storage sites

II Transport operations
- Unloading parts from carts to line-side standby storage site
- Transport of filled and empty parts containers is noncyclical
- Arrangement of filled and empty parts containers on carts

Figure 1-10. Comparison Between Current System and Synchronized and Equalized Production System

these days, we might go beyond the conservative 56 percent loss reduction figure and be bullish enough to claim a possible loss reduction of at least 60 percent.

Figure 1-11 illustrates the roles and assignments of semidirect operations after implementation of the proposed improvements. Compare this figure to Figure 1-4.

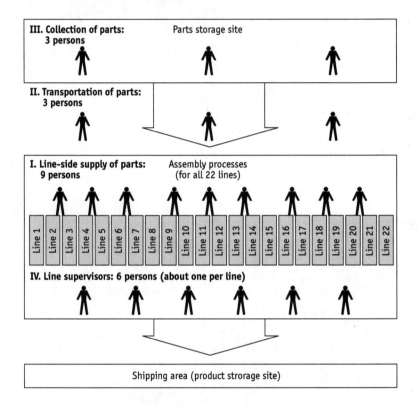

Basic approach to process management changes to "function-specific management."

There are 21 persons employed in indirect operations (semidirect operations). These include:

I.   Line-side parts supply staff   ...9 persons (including product transport staff)
II. Parts transport staff   ...3 persons
III. Parts collection staff   ...3 persons
IV. Line supervisors   ...6 persons

Figure 1-11. Overview of Roles and Assignments of Semidirect Operations Staff

Since all responsibility for production planning and execution belongs to the Planning Department, the semidirect operations staff function as implementation agents. Figure 1-12 shows an example of improvements in work methods that help to spell out this approach by listing the effects of work method improvements among line supervisors.

Work load ratio (percent)	Old system: Tasks performed by line supervisors		Structure of improvements						New system: Tasks performed by line supervisors
			Evaluation of work assignments (after making work improvements)						
	Task category	Description of work	Person at start of line	Line inspector	Parts collector	Planning Dept.	Eliminated tasks	Description of improvement (remarks)	
0									
10	Production indicators: 36%	Selection and distribution of parts configuration list and inspection drawings: 23%	Improvement No. 1					Automated search of parts configuration list and inspection drawings	
20					Improvement No. 2			Changed so that setup can be done by person at start of line	
30		Selection and distribution of product labels: 3% Stamping: 2% Issuing specifications: 1% Making arrangements: 7%	Improvement No. 3					Improved to enable handling of parts *Improvements in tools and jigs *Work-methods improvements	
40	Parts kanban: 32%	Issue parts kanban: 13% Processing of parts kanban 8%				Improvement No. 4		*No more making of arrangements, thanks to synchronization and equalization	
50		Processing of memos: 5% Confirmation of input sequence list, etc.: 4%							Issue parts kanban: 13%
60		Making arrangements and walking: 2%				Improvement No. 5		*This work is no longer necessary, thanks to synchronization and equalization	
70	Product kanban: 20%	Processing of product kanban checksheet: 7% Wrapping rubber bands around product kanban: 7%				Improvement No. 6		*Improvement in use of memos *This work is no longer necessary, since production is based on specifications from Planning Department	Confirmation of input sequence list, etc.: 6%
80		Setting up product kanban: 4% Filling in details about kanban: 2%	302 minutes (100 percent) Net work load per line supervisor per day • 169 minutes per person per day • (56 points ... 1/2 or less)			Improvement No. 7			Wrapping rubber bands around, setting up, and filling in details about kanban: 13%
90	Other tasks: 12%	Handling of other tasks: 12%	↓ 133 minutes (44 percent) Net work load per line supervisor per day						Handling of other tasks: 12 %
100									

Figure 1-12. Overview of Work Method Improvement Effects Among Line Supervisors

PHASE FOUR: FIVE STEPS FOR FLESHING OUT THE SYNCHRONIZED AND EQUALIZED PRODUCTION SYSTEM

As we carried out our studies, analyses, and improvement projects to flesh out our general ideas about a synchronized and equalized production system, we proceeded according to the five steps outlined below. Based on these five steps and with reference to examples from our consulting project, the following describes how we further developed the principles (discussed in the next chapter) behind a synchronized and equalized production system.

1. Clarify concepts behind a synchronized and equalized production system and present an overview of the system.
2. Establish a sequence for planning a synchronized and equalized production schedule.
3. Create a planning chart of equalized sizes, divisions of production specifications, and product inventory standards.
4. Establish an ESP Production pattern (basic master plan) for each line.
5. Confirm the operation of the ESP Production pattern.

Step 1: Clarify Concepts Behind a Synchronized and Equalized Production System and Present an Overview of the System

Figure 1-13 lists concepts and guarantees that we came up with based on the studies and analyses we had conducted at this point concerning a synchronized and equalized production system. These concepts and guarantees were the initial versions of what we would eventually call the Four Principles and Six Guarantees of the ESP Production System (discussed in Chapters 2 and 3). Figure 1-14 shows an overview of our synchronized and equalized production system. This is like a basic prototype of the current ESP Production System.

Step 2: Establish a Sequence for Planning a Synchronized and Equalized Production Schedule

We sought to establish a sequence for planning a synchronized and equalized production system in order to flesh out the concepts behind the synchronized and equalized production system so that we could actually operate such a system. We came up with three steps:

1. Zero missing items planning stage.
2. Overproduction prevention planning stage.
3. The load feed-forward planning stage.

Concepts behind the system
1. Guarantee against late deliveries and missing items. (Zero missing items plan)
2. Guarantee against excess inventories (products and parts). (Overproduction prevention plan)
3. Guarantee to anticipate and minimize production load fluctuations. (Load feed-forward plan)
4. Guarantee to enable a constant high operation rate. (Plan for a regular workday's maximum operation rate)

Figure 1-13. Initial Concepts and Guarantees Behind the Synchronized and Equalized Production System

These three steps would lead toward the operation of a system based on the *initial* four guarantees we had established for a synchronized and equalized production system. We also defined 15 steps among these three stages (see Figure 1-15).

Overview of the System

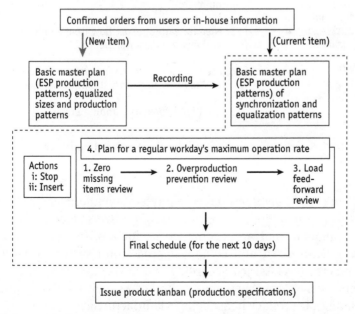

Figure 1-14. Overview of Synchronized and Equalized Production System

Plan for a regular work day's maximum operation rate

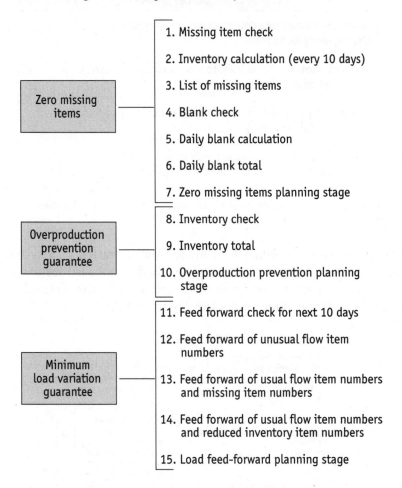

Figure 1-15. Steps in the Planning of a Synchronized and Equalized Production Schedule

The term *usual flow items* refers to product item numbers that are frequently in production, while *unusual flow items* refers to product item numbers that are seldom in production. In other words, under a synchronized and equalized production system, product item numbers are classified according to the size and frequency of the orders received for them, and this indicates their relative importance among all product items being manufactured.

Any item that is registered as an item that is usually in production as part of a basic master plan (ESP production pattern) is a usual flow item. (We define the *ESP production pattern* as an early, preliminary production plan that helps to clarify the company's production planning standards and anticipate demand from buyers.) Although an input sequence and production size (equalized unit) are set for unusual flow items, unusual flow items do not have their own production schedule in the ESP production pattern. In principle, after an order for an unusual flow item has been received, production of usual flow items is stopped so that the unusual flow item can be inserted into the production schedule. This approach of classifying product item numbers according to the size and frequency of their orders is extremely important. Under the fully developed, current version of the ESP Production System, usual flow items are called ESP highway items, which we will describe later.

The criterion for determining whether a product item number is a usual flow item or an unusual flow item is whether or not orders for the item are received basically every day. In the example of the Z Line described later (in Step 4), there are 36 product item numbers, of which 15 item numbers are usual flow items and 21 are unusual flow items. Figure 1-14 shows how the process for proposing a synchronized and equalized production schedule was established. This process clarified the functions and rules used to carry out the steps in proposing a synchronized and equalized production schedule. Figure 1-16 shows an overview of the process for proposing a synchronized and equalized production schedule.

Step 3: Create a Planning Chart of Equalized Sizes, Divisions of Production Specifications, and Product Inventory Standards

In order to establish a basic master plan—the ESP production pattern—for each of the 22 final assembly lines, we created a planning chart of equalized sizes, divisions of production specifications, and product inventory standards to serve as a set of standards (see Figure 1-17.) When creating this planning chart, we incorporated our prior research and analyses of orders received during the past couple of years, as well as trend predictions for each automobile manufacturer and car model.

In the example shown in Figure 1-17, six products can fit into each of the pass boxes that are used when delivering products to automobile manufacturers. That is why the equalized size is always a multiple of six. Also, in this example, the equalized sizes were set based on the amount of orders received (equal to the production

Plan for a regular work day's maximum operation rate

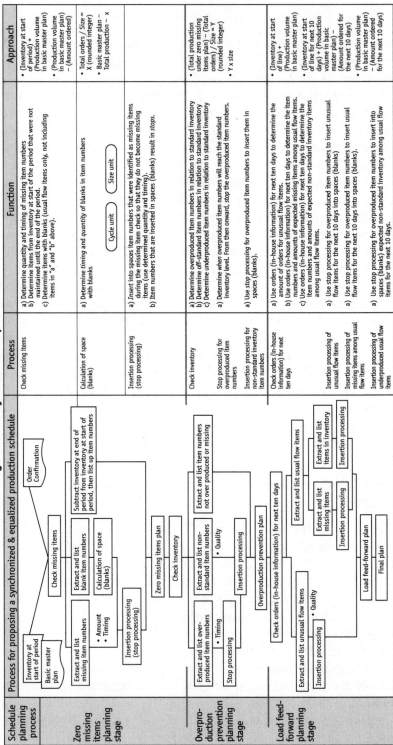

Schedule planning process	Process for proposing a synchronized & equalized production schedule	Process	Function	Approach
	(flowchart)	Check missing items	a) Determine quantity and timing of missing item numbers b) Determine items from inventory at start of the period that were not maintained until the end of the period. c) Determine items with blanks (usual flow items only, not including items in "a" and "b" above).	• (Inventory at start of period) + (Production volume in basic master plan) – (Amount ordered) • (Production volume in basic master plan) – x
Zero missing items planning stage		Calculation of space (blanks)	a) Determine *timing* and *quantity* of blanks in item numbers with blanks ⟨Cycle unit⟩ ⟨Size unit⟩	• Total orders / Size = X (rounded integer) • Basic master plan – Total production – x
		Insertion processing (stop processing)	a) *Insert* into spaces item numbers that were identified as missing items during the missing item check so that they do not become missing items (use determined quantity and timing). b) Item numbers that are inserted in spaces (blanks) result in *stops*.	
Overproduction prevention planning stage		Check inventory	a) Determine overproduced item numbers in relation to standard inventory b) Determine off-standard item numbers in relation to standard inventory c) Determine underproduced item numbers in relation to standard inventory	• (Total production under zero missing items plan) – (Total orders) / Size = Y (rounded integer) • Y x size
		Stop processing for overproduced item numbers	a) Determine *when* overproduced item numbers will reach the standard inventory level. From then onward, *stop* the overproduced item numbers.	
		Insertion processing for non-standard inventory item numbers	a) Use *stop processing* for overproduced item numbers to *insert* them in spaces (blanks).	
Load feed-forward planning stage		Check orders (in-house information) for next ten days	a) Use orders (in-house information) for next ten days to determine the amount of orders for unusual flow items. b) Use orders (in-house information) for next ten days to determine the item numbers and amounts of expected missing items among usual flow items. c) Use orders (in-house information) for next ten days to determine the item numbers and amounts of expected non-standard inventory items among usual flow items.	• (Inventory at start of line) + (Production volume in basic master plan) – (Amount ordered for the next 10 days) • (Inventory at start of line for next 10 days) + (Production volume in basic master plan) – (Amount ordered for the next 10 days) • (Production volume in basic master plan) – (Amount ordered for the next 10 days)
		Insertion processing of unusual flow items	a) Use stop processing for overproduced item numbers to insert unusual flow items for the next 10 days into spaces (blanks)	
		Insertion processing of missing items among usual flow items	a) Use stop processing for overproduced item numbers to insert usual flow items for the next 10 days in spaces (blanks).	
		Insertion processing of underproduced usual flow items	a) Use stop processing for overproduced item numbers to insert into spaces (blanks) expected non-standard inventory among usual flow items for the next 10 days.	

Figure 1-16. Overview of Process for Proposing a Synchronized and Equalized Production Schedule

volume) per month. In Figure 1-17, the column heading *Standard for number of times specifications are issued (out of four cycles per day total)* refers to the number of times per day production specifications are issued for a particular product item, out of the four cycles per day (once every two hours) that production specifications are issued. This standard is used to determine how requirements for issuing specifications will be incorporated into the ESP production pattern.

This is an important standard, not only for use in establishing a plan for a regular workday's maximum operation rate at this supplier's final assembly processes, in which the amount of production specified in the production specifications issued every two hours equals 100 percent of the production capacity, but also for use in computer simulations. In this standard, since there are no differences among the assembly labor hours for different product item numbers, the production load can be converted into units.

In other words, the production specifications that are issued every two hours must be constant in terms of the resulting production load. To accomplish this, you combine the product item numbers based on the standard for the number of times specifications are issued (out of four cycles per day total) so that you can incorporate the requirement of a constant production volume into the ESP production pattern (basic master plan).

The column heading *No. of days of inventory (standard)* is the standard for limited inventory levels. You use the number of days of inventory as the criterion for determining when to execute stops and insertions when drafting the zero missing items plan, the overproduction prevention pan, the load feed-forward plan, and the plan for a regular workday's maximum operation rate. (Stops and insertions are discussed in Chapter 3.)

Step 4: Establish an ESP Production Pattern for Each Line

Finally, we were ready to establish an ESP production pattern (basic master plan) for each line. (We will discuss the ESP production pattern in further detail in Chapter 3.) Let us examine an example: the ESP Production pattern for the Z Line shown in Figure 1-18.

Thus, for the Z Line, there are 36 product item numbers, and production specifications for 180 units issued every two hours (or 720 units per regular 8-hour working day) to achieve 100 percent of the production capacity. The ESP Production pattern in Figure 1-18

Production volume per month	Equalized size	Standard for number of times specifications are issued (out of four cycles per day total)	No. of days of inventory (standard)
9,501 ~10,000	96		
9,001 ~ 9,500	96		
8,501 ~ 9,000	90		
8,001 ~ 8,500	84		
7,501 ~ 8,000	78		
7,001 ~ 7,500	72		
6,501 ~ 7,000	66		
6,001 ~ 6,500	66		1.5 days
5,501 ~ 6,000	60	A (four cycles)	
5,001 ~ 5,500	54		
4,501 ~ 5,000	48		
4,001 ~ 4,500	42		
3,501 ~ 4,000	36		
3,000 ~ 3,500	36		
2,501 ~ 3,000	30		
2,001 ~ 2,500	24		
1,501 ~ 2,000	18		2.0 days
1,001 ~ 1,500	18		
501 ~ 1,000	12	B (three cycles)	2.5 days
201 ~ 500	12	C (two cycles)	3.0 days
101 ~ 200	6	D (one cycle)	4.0 days
1 ~ 100	6		5.0 days

Figure 1-17. Planning Chart of Equalized Sizes, Divisions of Production Specifications, and Product Inventory Standards (Example)

shows an input sequence (numbered 1 to 36) for item numbers A to jj. Note that the input sequence is divided into two parts, the first part being for usual flow items and the second for unusual flow items. The input sequence for this final assembly line is organized to facilitate quality assurance (i.e., prevention of incorrect parts, missing parts, or incorrectly assembled products).

A two-hour pattern appears in the four cycle columns (Cycle 1 to Cycle 4). The item numbers and quantities shown in regular type in the shaded cells were set as part of the production pattern. The "blank cells" shown with bold letters are not part of the production schedule in the ESP production pattern. However, a production schedule value of zero is entered in each of these blank cells to indicate that they are available for insertions.

The total of usual flow items cells indicates that 180 units are produced every two hours, which is the requirement to be met in the Z Line's plan for a regular workday's maximum operation rate. Whenever unusual flow items are produced, the production of usual flow items is stopped so that an unusual flow item can be inserted to the available slot (blank). Naturally, if for some reason you cannot complete the ESP production pattern within the required period, such as when you receive an overload of orders, you must schedule overtime work or work on weekends or holidays to make up the difference.

Step 5: Confirm the Operation of the ESP Production Pattern

After building a synchronized and equalized production system for the Z Line, we confirmed its feasibility by comparing the production schedule (production specifications) before and after the introduction of the synchronized and equalized production system. Figure 1-19 shows some examples of these before-and-after comparisons.

As shown in the figure, before making this improvement the number of items produced varied widely from day to day and from cycle to cycle, and introducing a synchronized and equalized production system effectively eliminated this wide variation. In the bottom chart (for item number A), the gaps indicated by the diagonal lines indicate times when the equalized amounts or sizes were revised based on in-house information. This demonstrates the importance of ensuring reliability by revising (maintaining) the ESP production pattern (basic master plan) according to the latest received order data and in-house information. Such revisions provide assurance that you will operate the synchronized and equalized production system using accurate data and clear standards. This

Z Line (ESP Production Pattern)									
Category		Cycle 1		Cycle 2		Cycle 3		Cycle 4	
		Item #	Production size	Item #	Production size	Item #	Production size	Item #	Production size
Usual flow items	1	A	30	A	30	A	30	A	30
	2	B	30	B	30	B	30	B	30
	3	C	18	C	18	C	18	C	18
	4	D	12	D	12	*(D)*	*0*	*(D)*	*0*
	5	*(E)*	*0*	*(E)*	*0*	E	12	E	12
	6	F	12	F	12	*(F)*	*0*	*(F)*	*0*
	7	*(G)*	*0*	*(G)*	*0*	G	12	G	12
	8	H	12	H	12	H	12	*H*	*0*
	9	I	18	I	18	I	18	I	18
	10	J	12	J	12	*(J)*	*0*	J	12
	11	*(K)*	*0*	*(K)*	*0*	*(K)*	*0*	K	6
	12	L	12	*(L)*	*0*	L	12	L	12
	13	*(M)*	*0*	*(M)*	*0*	*(M)*	*0*	M	6
	14	*(N)*	*0*	N	12	N	12	*(N)*	*0*
	15	O	24	O	24	O	24	O	24
Total of usual flow items =			180		180		180		180
Units required for 100 percent production rate per cycle (every two hours)									
Unusual flow items	16	P	6	P	6	P	6	P	6
	17	Q	6	Q	6	Q	6	Q	6
	18	R	6	R	6	R	6	R	6
	19	S	6	S	6	S	6	S	6
	20	T	6	T	6	T	6	T	6
	21	U	6	U	6	U	6	U	6
	22	V	6	V	6	V	6	V	6
	23	W	6	W	6	W	6	W	6
	24	X	6	X	6	X	6	X	6
	25	Y	6	Y	6	Y	6	Y	6
	26	Z	6	Z	6	Z	6	Z	6
	27	aa	6	aa	6	aa	6	aa	6
	28	bb	6	bb	6	bb	6	bb	6
	29	cc	6	cc	6	cc	6	cc	6
	30	dd	6	dd	6	dd	6	dd	6
	31	ee	6	ee	6	ee	6	ee	6
	32	ff	6	ff	6	ff	6	ff	6
	33	gg	6	gg	6	gg	6	gg	6
	34	hh	6	hh	6	hh	6	hh	6
	35	ii	6	ii	6	ii	6	ii	6
	36	jj	6	jj	6	jj	6	jj	6

Input sequence

Figure 1-18. Example of ESP Production Pattern (Basic Master Plan) for "Z Line"

updating of information for revisions (maintenance) is an area that lends itself to automation (computerization).

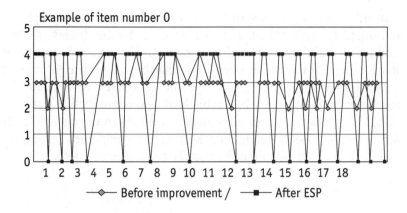

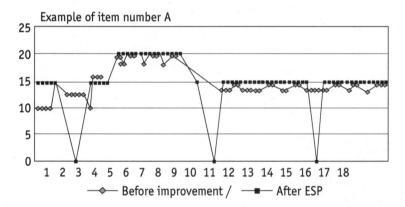

Figure 1-19. Example of Item-Specific Production Schedules Before and After Introduction of Synchronized and Equalized Production System

CONCLUSION—THE BIRTH OF A NEW PRODUCTION SYSTEM

This concludes the case study of this automotive electrical components manufacturer that became the springboard for our subsequent development of the ESP Production System. This brief recounting of this seminal case study should help the reader to better understand the essence of the ESP Production System, as well as assist the reader as he or she progresses through the remainder of the book. It is important to reiterate here what we ultimately learned and realized while working on this consulting project: while the JIT production system (using kanban) is quite suitable for end buyers such as automobile manufacturers, it does not necessarily work so well for the suppliers that deliver components to these buyers.

In other words, we discovered that JIT is not for everyone and that suppliers will continue to suffer considerable losses in their production activities unless they can establish a production system that is suited to their own needs. So what began as straightforward consulting assignment to increase the efficiency of a supplier's assembly lines, ended up in the birth of a new production system— ESP. It was conceived and ultimately developed to work specifically and powerfully as a production system for companies (especially suppliers) that are ill suited for the JIT production system.

Why Suppliers Need the ESP Production System

Today, the Just-in-Time (JIT) concept (purchase and produce just what is needed, just when it is needed, and in just the needed amount) and, especially, its role in the Toyota Production System, has gained worldwide recognition and praise. In fact, so many manufacturing companies have implemented JIT that one might reasonably regard it as an industry of its own. Nevertheless, the JIT production system is not the best choice for all manufacturers. In fact, many manufacturing suppliers suffer in their attempt to adhere to the strict requirements imposed on them by the JIT production systems of their buyers. In this chapter, we will discuss what we see as the exploitative practices resulting from this supplier-buyer relationship and the causal factors. In the next few chapters, we use the term *buyer* to refer to the type of company that processes and/or assembles final products to be shipped to consumers. The term *supplier* is used to refer to a manufacturing company that delivers parts and/or materials to a buyer. In the case studies later on, we will use the term *customer* rather than *buyer*.

ROOTS OF THE ESP PRODUCTION SYSTEM

The ESP Production System is the name of a program presented by JMAC to implement a synchronized mass-production system. ESP stands for "Equalized and Synchronized Production." It was developed as a solution for those manufacturing companies or suppliers that turn out a continuous repeated flow of products and that would have or already have had problems implementing the JIT production system.

When a company is ill-suited for or unsuccessful with implementing JIT, it can lead to lots of problems or other surprisingly poor results. For example, the company may find it cannot reach its goal of reducing inventory, such as products, in-process parts, and other parts and materials. To the contrary, they find themselves generating dead inventory. The company's Production Management Division becomes preoccupied with delivery management, making it more difficult to generate accurate production specifications and purchase parts and materials. Meanwhile, the Manufacturing Division is scrambling to fulfill existing production specifications while dealing with the frequent changes due to last-minute additions and/or cancellations. As a result, it becomes impossible to maintain high productivity among both employees and equipment, resulting in various problems such as inefficiency and having to work overtime or on holidays.

Companies that have experienced these types of problems are either in an industry or business category that is ill-suited for JIT, or they've misunderstood its basic nature, such as when they pay attention only to JIT's operational aspects or have imitated its form but not its substance. Many of these superficial imitators of JIT are manufacturing companies that have limited their focus to one operational aspect, namely the *kanban system*. When a company operates kanban in this way, it is referred to as the *imitation kanban system*.

In many cases, companies that are ill-suited for JIT are manufacturing companies, or suppliers, whose buyers (final destination companies or factories) use the JIT production system. As a result, these suppliers are obliged to suffer the various drawbacks of JIT to meet their buyer's strict requirements. In other words, because buyers demand that their suppliers adhere and adjust to their JIT programs, the supplier ends up being exploited by it (see Figure 2-1). This exploitation becomes obvious when a supplier must deliver products according to the various JIT-generated requirements of several buyers at the same time or when it has to deal with several buyers operating an imitation kanban system. Thus various problems can occur as a result of the exploitative practices of the JIT production system, such as:

1. It often happens that a supplier must scramble to make quick deliveries in response to unreasonable requests from buyers just to keep these buyers' own production schedules operating at full speed.

2. The supplier must add a much greater variety of products and yields in order to coordinate deliveries with the buyer's own production schedule. Such variety increases exponentially when the coordination of production must be implemented for several buyers at the same time. Under such conditions, the supplier can give little or no consideration to the production efficiency of their own production lines.

3. To meet the buyer's requirements, the supplier must maintain production processes that allow for a certain *availability rate*. (*Note*: The availability rate is the indicator of how available a manufacturer is to responding immediately to a production request.) Production processes that preserve an availability rate typically require the use of production lines that are dedicated to specific buyers, which in many cases requires the supplier to invest in production equipment without regard to the investment's cost-effectiveness.

4. The basic idea behind having production lines dedicated to specific customers is to enable production to be more easily adjusted to meet the orders of specific customers. This means that the fluctuation in orders from a customer are directly reflected in the amount of production dedicated to that customer, and there is little the supplier can do to maximize in-house production efficiency. Often, the upshot is that the supplier must deal with a situation in which both labor and equipment resources are wasted.

5. To meet buyers requirements for JIT deliveries, the supplier must either locate the final production process for that buyer close to the buyer's own facilities or must establish some kind of warehouse to keep products close to them. In either case, the supplier is obliged to invest in resources that may not be cost-effective.

6. In cases where the supplier is unable to establish production processes or a warehouse close to a certain buyer, the typical solution is to use trucks to deliver products to distant buyers according to their JIT schedules. Problems encountered by suppliers who rely on trucks to make JIT deliveries include:

   • Having to maintain a large fleet of dozens of trucks to meet the buyers' requirements for frequent deliveries.

   • Having to make deliveries in various quantities (as per the buyer's order) resulting in trucks seldom being filled to capacity. Trucking becomes less cost-effective.

   • Creating congested highways and expressways due to having to make frequent truck deliveries among many buyers, thereby lowering transportation efficiency (due to stop-and-go traffic) and worsening traffic jams near buyers' sites.

Figure 2-1. Exploitive Practices Imposed on Suppliers by the Buyer's JIT Program

- Working out the differing responsibilities with the various buyer relationships.
- Having to satisfy several highly demanding buyers.
- Having to rely overly on a buyer's imitation kanban system.

In the remainder of this chapter, we will explain these underlying problems and how ESP addresses them.

FOUR PRINCIPLES BEHIND THE CONSTRUCTION AND OPERATION OF ESP

As we have noted, ESP is a solution for companies that are ill-suited for or are being exploited by their buyer's JIT Production System. ESP uses synchronized and equalized production methods to then realize the Six Guarantees that will eliminate the pitfalls of JIT (see Figure 2-2). As will be further discussed in Chapter 3, these six guarantees are the basis for successfully implementing and maintaining the ESP Production System. But before we address these guarantees, you need to understand the Four Principles behind the construction and operation of ESP (see Figure 2-3). Without these four principles you cannot achieve ESP's Six Guarantees.

1. The zero missing products guarantee	Zero late deliveries
2. The overproduction prevention guarantee	Zero excess inventory (zero or minimal inventory)
3. Production load feed-forward guarantee	Production loads are systematically coordinated to minimize day-to-day fluctuations
4. Production efficiency maximization guarantee	Schedule and production seek to maximize operating rate of production processes
5. Production change adapability guarantee	Flexible responses to additions and cancellations
6. Logical production management guarantee	Thorough use of computers in production management, thorough rejection of obsolete manager-based decision making

Figure 2-2. The Six Guarantees of the ESP Production System

1. Recognizing that productivity is chiefly determined at the production planning stage, it is therefore best optimized at the production planning stage.

2. Control of inventory (products, in-process goods, parts, materials) is best determined at the production planning stage.

3. There is no point in constructing an ESP Production System that does not include improvement of production processes.

4. The Production Management Division must lead (by setting targets and goals and evaluating implementation) the improvement of production processes.

Figure 2-3. Four Principles Behind the Construction and Operation of ESP

Principle One

The most basic principle is to proceed as if everything is decided at the planning stage. Consequently, ESP's greatest emphasis is on intensifying functions and managing operations at the planning stage. This is an absolutely essential principle that guides all activities under the ESP Production System. Moreover, it is this principle that is the key to achieving, maintaining, and expanding the Six Guarantees of ESP.

Recognizing that productivity (i.e., the basic level of productivity) is mainly determined at the production planning stage, it is therefore best to optimize it at the production planning stage. Under ESP, it is an absolute requirement that during the production planning stage you work the hardest to minimize loss. Examples of loss include:

- **Loss due to delays in planning and management loss.** When the production load fluctuates or when overproduction occurs.
- **Line stoppage loss.** Combining or switching products on the production line in ways that lower production efficiency.
- **Defect loss.** Combining products on the production line in ways that tend to raise the defect rate.

The structure of production efficiency is generally similar to that shown in Figure 2-4.

First and foremost, ESP focuses on planning delay loss and management loss, which is the differential between the total shift time

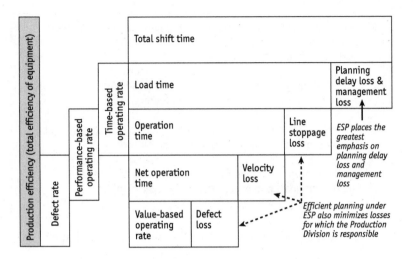

Figure 2-4. The Basic Level of Production Efficiency Is Determined at the Production Planning Stage

and the load time. Ordinarily, you can define this planning delay loss and management loss as the paused time during which the Production Division must temporarily stop production operations due to a gap in the production schedule. For example, a processing or assembly line or other production equipment may have to undergo a scheduled stop in order to coordinate production, since there is no production scheduled at the moment. Or, you may have to stop a processing or assembly line or other production equipment to wait on standby when certain parts or materials are missing. Production lines are also stopped during certain assemblies, meetings, and improvement activities, and all of these stoppages are included in this concept of stopped or paused production.

Except in cases where you have to adjust production due to poor sales, you can generally trace the causes of planning delay loss and management loss to vague planning standards. Such losses tend to occur when the planning standards are determined on results and reports from the Manufacturing Division rather than the Production Management Division. Or losses may still occur even when the Production Management Division has established initial planning standards, but their revision, as part of maintenance, is neglected when subsequent changes or improvements are made in the product structure and assembly methods.

Production planning using an inner circle

Within any company's Production Management Division, people tend to think first of all about producing the products that have been ordered. Managers and others in the Production Management Division tend to reason, if we can just get the specifications out to the Production Division to produce the amount based on the current orders, we can take care of everything else later. Or they may think, the production staff can work overtime and holidays when we have an overload of orders, and they can make up for it later when the production load is low.

At some companies, various planning standards are regarded as something that the Production Management Division (an indirect division) and the Manufacturing Division (a direct division) negotiate. In such cases, if the staff in the Production Management Division do not heed the advice of their counterparts in the Manufacturing Division, they will likely find it hard to deliver products to buyers on time and will also tend to fall short of their own production management targets in terms of cost management, profit management, quality management, and so on. When you leave individual planners to determine and work out the planning standards with their counterparts in the Manufacturing Division, there can be, due to the lack of transparency, some degree of uncertainty resulting with these planning standards. The end result is the creation of an inner circle of people who carry out the production management work.

ESP creates clear planning standards

Under the ESP Production System, you establish clear planning standards so that right from the production planning stage, you can make every effort to schedule the maximum possible production output (i.e., the maximum production load within the limits of the production capacity). You can rationally configure all production management tasks and define each step, which will eliminate the old inner circle approach to production planning. Also, when changing the product structures or processing/assembly methods, or when making improvements in the production processes, you can immediately review and revise (maintain) the planning standards. Consequently, under ESP, the Production Management Division takes the leading role in improving manufacturing processes.

However, this does not mean that the production managers are free to ignore whatever input or independent-minded ingenuity may come from the Manufacturing Division.

When proposing a production schedule, the Production Management Division can work most efficiently by gaining a grasp of which production processes are likely to involve obstacles, such as bottlenecks caused by short lead time, shorter changeover times needed to produce smaller lots, and quality improvements needed to increase production yield. Then they can spell out to the Manufacturing Division what improvements to make and targets to reach, overseeing the implementation phase, and finally checking the results and capturing any lessons from them in the future planning standards.

When using this approach, it becomes possible to gain an overall view of the production processes and to devise specifications that will help reach the goal of overall optimization. To do this, the Production Management Division must be allowed to exercise the needed regulatory functions. It bears repeating that none of this should excuse or preclude the Manufacturing Division staff from independently devising improvements. Instead, it simply means that the Manufacturing Division should give the highest priority to improvements aimed at overall optimization proposed by the Production Management Division, before implementing those improvements independently devised by the Manufacturing Division. The Production Management Division should be responsible for thoroughly understanding all production processes and devising efficient production plans based on that understanding. By gaining a detailed grasp of the Manufacturing Division's production requirements, the Production Management Division can come up with a production schedule that minimizes the line stoppage loss, velocity loss, and defect loss mentioned in Figure 2-4.

One clear example of the rule of having a detailed grasp of production requirements concerns changeover and adjustment loss, a type of line stoppage loss that can be avoided by setting the appropriate sequence of production processes. One might set a rule that requires large products to be handled before small products (or vice versa) at a certain machining process, or that requires you to arrange production processes so as to minimize the number of jig and tool replacements. Another proposed rule might be that lighter colors are done before darker colors at coloring processes, such as when manufacturing pigments or color-coating products. Yet another rule might be that wide pieces come before narrow pieces at a metal rolling process. A production schedule that observes all of these rules would be a schedule that works to minimize line stoppage losses, particularly changeover and adjustment losses.

By way of addressing defect loss in an assembly process, one might first determine which sequence of production processes will work best in preventing selection and use of incorrect assembly parts, then propose a production schedule that uses the determined sequence. The idea, therefore, is to establish an input sequence (production sequence) that is clearly based on the principle of improving the efficiency of the company's production processes. If you can devise a production schedule that incorporates such an input sequence, you will minimize line stoppage loss, such as occurs from changeovers, adjustments, and breakdowns, velocity loss, such as occurs when standard times are not (or cannot be) maintained, and defect loss caused by the lowering of product quality and production yield.

Principle Two

The second principle behind the construction and operation of ESP is that control of inventory (products, in-process goods, parts, and materials) is best determined at the production planning stage. This principle suggests that the Production Management Division does not bear the primary responsibility for controlling various levels (types and amounts) of inventory, such as products, in-process goods, parts, and materials. Instead, the Production Management Division, after matching planned production to orders received and after carefully analyzing the Manufacturing Division's production capacity, proposes a production schedule that is designed for maximum production efficiency. Next, it carries out timely purchases of the parts and materials needed for that production schedule. If it can do all of that, the Production Management Division can control the entire inventory and has only itself to blame if it cannot effectively manage inventory (including estimating inventory and checking actual inventory levels).

Thus, the approach under ESP is that the Production Management Division is responsible for both setting and achieving basic levels of production efficiency and inventory, while the Manufacturing Division plays only a limited role in those areas. After all, production efficiency and inventory are not results that come from the implementation of production activities, nor are they determined by some kind of natural, spontaneous process. Both production efficiency and inventory have a rational structure and are areas that you must manage by statistical control. Under ESP the Production Management Division must stick to this principle. In other words, they must envision their intended results and then strive to achieve actual results in line with the intended results.

Principle Three

The third principle is that there is no point in constructing an ESP Production System that does not include improvement of production processes. This means that significant improvement of production efficiency is not possible unless you improve the manufacturing processes, or the production system's hardware. But if bottlenecks or other constraints in production processes are left as they are, it will be impossible to establish a truly good production system or achieve truly good results, no matter how many improvements are made in the production management system, or the production system's software.

In sum, ESP seeks to realize the maximum possible efficiency in production functions overall by maximizing efficiency in both manufacturing processes (hardware) and the production management system (software). It does this by finding ways to use all production resources, like the 4Ms (man, machine, material, and methods), as efficiently as possible while also implementing improvements that cut costs and enhance profitability.

The overall objective of manufacturing process improvements under ESP is to be able to respond effectively to any order from any buyer while also raising in-house production efficiency and minimizing inventory. Key strategies to achieve this objective include thoroughgoing efforts to shorten lead times and use smaller lots combined with a substantial improvement in production capacity (see Figure 2-5).

As described above, shortening lead time, using smaller lots and raising in-house production efficiency are all key activities under ESP. These three achievements will enable the company to respond fully to all buyer orders and to create manufacturing processes that help maximize in-house production efficiency while minimizing inventory. The combination of shortening lead times and using smaller lots will markedly improve the company's abil-

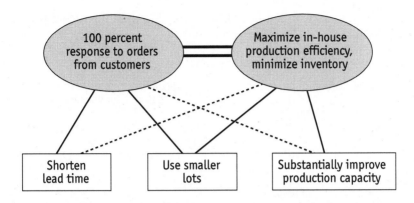

Figure 2-5. Objectives of Manufacturing Process Improvements Under ESP

ity to respond to orders from buyers. Also, after shifting to the use of smaller lots, if the company can also substantially boost production capacity, there will be a substantial payoff in terms of flexibility in production scheduling. This in turn will make it easier to respond fully to buyer orders. Naturally, it also facilitates inventory reduction.

Four steps to achieving manufacturing process improvements

There are four steps for achieving the manufacturing process improvements that will transform current manufacturing processes into a dream come true (see Figure 2-6). The approach that underlies these four steps involves thoroughly applying the ECRS principles of improvement during implementation of the four steps. ECRS is an acronym that stands for:

- Eliminate
- Combine
- Rearrange
- Simplify

The order of the ECRS principles is based on a scale of effectiveness, so you should implement actions to make improvements based on this order. The most effective type of improvement is achieved through the action of eliminating. Then comes the action of combining or rearranging, while simplifying tends to bring about the lowest level of improvement. ESP requires faithful adherence to these ECRS principles while working to improve manufacturing processes. In fact, it can be said that making manufacturing process improvements via this approach is indispensable, both for establishing ESP and for achieving its promised results.

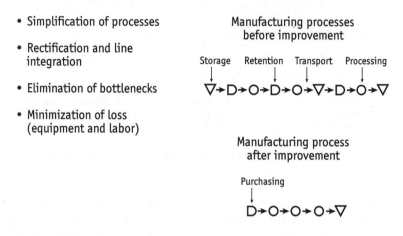

- Simplification of processes

- Rectification and line integration

- Elimination of bottlenecks

- Minimization of loss (equipment and labor)

Figure 2-6. Steps in Achieving Manufacturing Process Improvements Under ESP

Principle Four

The fourth principle behind ESP is that the Production Management Division must lead (by setting targets and goals and evaluating implementation) the improvement of production processes. This principle underscores how the role of the Production Management Division is more than that of a Production Control Division (i.e., to propose plans and coordinate feedback concerning results). Its role as a full-fledged Production Management Division is to act as the flagship of the company's entire fleet, that is, its entire set of production functions. Without the Production Management Division there to guide them, the other ships (such as the Manufacturing Division), no matter how efficient or large they might be, will sail aimlessly without ever reaching their destined shore.

To build manufacturing processes that respond swiftly to the ever-changing needs of buyers, the Production Management Division must continually monitor current manufacturing conditions and devise effective improvements while in production. They also need to follow up on the company's performance and revise the next set of planning standards accordingly.

Figure 2-7 conceptually diagrams what we've discussed so far about ESP and shows that improving the manufacturing processes is the foundation for achieving ESP's Six Guarantees listed in Figure 2-2. Using the techniques of synchronized and equalized production, the Production Management Division can use effective planning as its core to become the manufacturing flagship that drives the success of ESP.

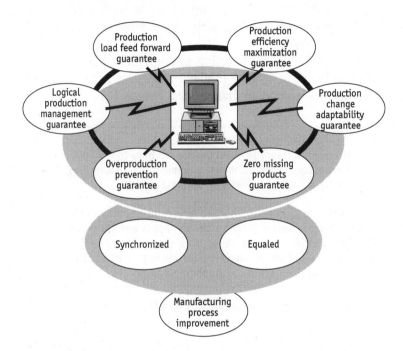

Figure 2-7. Conceptual Diagram of the ESP Production System

COMPANIES BEST SUITED FOR ESP

JMAC developed the ESP Production System to solve the problems manufacturing companies are having because they are either ill-suited for the JIT Production System or are being exploited by the buyer's kanban system. More specifically, it is an answer for manufacturing companies that are searching for a production system that serves their needs rather than struggling to adhere to the demands of JIT. As such, ESP, unlike JIT, is not designed for all types of industries and companies. But rather than to make a side-by-side comparison of the relative strengths and weaknesses of ESP and JIT, it is more useful to compare them in terms of the types of companies that are best suited for both production systems.

Basically, ESP works best for manufacturing companies that are suppliers who turn out a continuous, repeated flow of products, or who deliver semifinished products to final destination companies (buyers) that perform the final processing and/or assembly work. It is a production system that should be considered by manufacturing managers who are thinking, "We introduced JIT, but somehow it has not been working well," or "I wonder if JIT will really suit our company?"

For example, ESP will function as something like an escort ship and pilot ship for a supplier that manufactures parts used in automobiles and household appliances (see Figure 2-8).

---

1. **ESP** is typically more appropriate for a "parts supplier" that delivers products made on its production lines to multiple buyers or to one buyer that has multiple delivery sites. Either way, products are being delivered to several locations.

2. **JIT** is typically more appropriate for a company that purchases parts and/or materials from a parts supplier and assembles finished products, or a parts supplier that has built dedicated production lines in which all production is synchronized with the production lines of the corresponding buyers.

Figure 2-8. Suitable Companies for ESP and JIT (Example 1)

Another example where ESP has recently been very effective is at a company that produces food products, medical supplies, and precision-machined products to a national chain of stores and sales outlets (see Figure 2-9).

From the perspective of these two examples, we can say that companies where you could effectively introduce ESP would include the following conditions.

- An order production type of company that manufactures products in a continuous, repeated flow.
- A company that is being exploited as it strives to meet orders sent from a parent company, multiple buyers, or sales agencies (outlets).
- A company that has had little success in previous trial-and-error efforts to establish true wide-variety, small-lot production.

Generally, most companies that are best suited for ESP fall into the category of suppliers that deliver parts and/or materials to a

Companies being exploited by a manufacturer of finished products

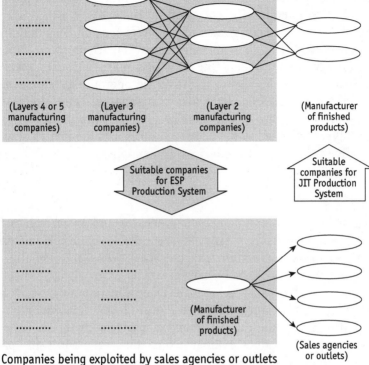

Figure 2-9. Suitable Companies for ESP and JIT (Example 2)

buyer. The author strongly suggests that, in any case, readers should gain an understanding of ESP and then determine whether or not it is appropriate for their companies.

HOW JIT AND KANBAN EXPLOIT SUPPLIERS

The first question is, is it really a good idea to plan your production system according to the demands of your buyers? As you will see, there are differing rationales of the buyer and the supplier. On one hand, buyers must focus primarily on their own needs for raising production efficiency or reducing inventory, which usually means that not only are they indifferent to their suppliers' similar needs, but expect their suppliers to adapt to their own production schedule for ordering materials and parts. On the other hand, suppliers must find ways to adjust their production and inventory to deliver on the orders of the buyer. Thus, right from the beginning, there are conflicting rationales about how buyers and suppliers approach managing production efficiency and inventory.

ESP is based on the rationale of suppliers, since it is usually these companies that struggle with adhering to the JIT program imposed on them by their buyers. One of ESP's main purposes includes establishing various ways to improve the operational management of suppliers. At its core lies the idea that suppliers cannot raise their production efficiency or cut their inventory levels if they only do what buyers tell them to do. Let us examine a little more closely the respective rationales of the buyers and suppliers.

Rationale of Buyers

Since a buyer is a company that processes and/or assembles finished products, it is very strict about ordering parts and materials according to its production schedule. Also, when ordering from its suppliers, the buyer is very careful to maintain an efficient cash flow so it will order only what it needs, in the amount it needs, and at the required delivery time (see Figure 2-10). Consequently, the buyer orders the minimum number of parts or materials for each product type or product number (using the smallest lots and packages possible), while specifying a particular date and time for delivery. It also places separate orders for the production lines and/or processes at the delivery destinations (see Figure 2-11.)

To maintain little or no inventory, a buyer is also very careful about the timing of its orders. It will not order parts and materials from suppliers until the final data is in concerning its ordered finished products and/or sales trends. This way the supplier's deliveries will arrive just in time for the start of production (see Figure 2-12.) Thus, the buyer does everything it can to minimize its inventories of finished products, in-process goods, parts, and materials.

When a buyer uses JIT, it makes effective use of the kanban system to make the process of ordering less cumbersome while also facilitating inventory management. In addition to using kanban, a buyer might implement a sequential delivery system in which they synchronize the schedule for the delivery of parts and materials according to the order in which they are used for the production of finished products. This helps to minimize inventory levels (especially under a zero-inventory campaign) of

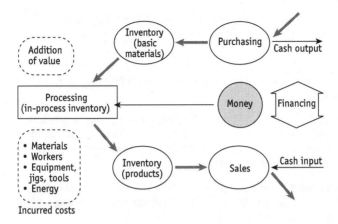

1. Funds are needed from the point where payment is made for materials to the point where payment is received for sold products. This is affected by the production period.

2. An excess of in-process (or warehouse) inventory requires a proportional excess of funds.

3. When warehouse inventory and in-process inventory are reduced, the production period can be shortened and less funds are tied up, which adds up to a more efficient cash flow.

Figure 2-10. Cash Flow with Production Timing and Inventory

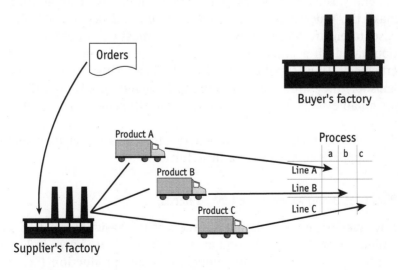

Figure 2-11. Needs of Buyer

parts and materials, especially those that take up a lot of storage space or tie up a lot of money.

Example of a Japanese buyer's timing when ordering parts and materials

• Kanban system ... There are 32 deliveries per day. A new order goes out after 16 deliveries (after 16 kanban are issued). In other words, parts and materials are ordered one-half day before they are to be delivered.

• Sequential delivery system ... Parts and materials are ordered 30 minutes before they are to be input into the processing and/or assembly line for finished products.

Figure 2-12. Example of Buyer's Timing when Ordering Parts and Materials

Rationale of Suppliers

Since a supplier is a company that delivers parts and materials that a buyer uses to process finished products, its production and delivery schedules are based on the orders it receives from its buyers. However, usually a supplier can't wait until they receive an order via a kanban or sequential delivery system (such as was shown in Figure 2-12) before starting production of the ordered parts and/or materials. Otherwise, a supplier would seldom have enough time to make the ordered goods and meet the buyer's delivery deadlines. Also, to ensure profitability, the supplier must carry out its production based on a production schedule that enables high production efficiency. Thus, suppliers are typically faced with one of two problems: (1) they are unable to meet their buyers' needs, and (2) they are unable to achieve high production efficiency. In this regard, there is a clear conflict between the rationales of the buyers and suppliers. In fact, they are diametrically opposed. Let us examine the supplier's predicament a little further.

Pitfalls of trying to meet the buyer's needs

If we take the buyer's timing of orders as the basic information signaling the start of production at the supplier, as is shown in Figure 2-12, there is usually not enough time for production (i.e., not enough net production lead time). But even if there were enough time for production for a particular product or supplier company, it would still be extremely difficult for the supplier to obtain or adjust the production resources in a timely manner for each type of product that was ordered. This would include purchasing materials, allocating time on production equipment, and allocating labor. After all, as shown in Figure 2-12, the actual job of handling orders involves receiving many orders per day in which there is great deal of variation concerning the specified parts and materials. The ordered parts and materials vary not only in type, but also in amount per model and total amount, so it is no easy task for the supplier company to keep track of all the orders it receives.

Generally, suppliers have mitigated such difficulties by adopting a JIT Production System that offers tools such as JIT-based coordination of manufacturing or a production control system such as the kanban system (see Figure 2-13). As for JIT-based coordination of manufacturing, this typically involves creating production lines that are dedicated to particular buyers (or delivery sites) and/or particular product types and allocating production resources accordingly. Once these specialized lines are in place, it becomes easier to absorb variation in ordered goods by spreading it out among specific products.

In theory, having a flexible production line that is able to turn out a variety of products would help, but it is very difficult to design flexible lines that can accommodate the wide variety of production variables that are involved, including processing and assembly methods and processing and assembly times. It would also require a huge investment. When one also considers the time and money that are needed to make a production line that can be promptly altered to meet product design changes, it is no wonder that no manufacturing company has managed to build a flexible manufacturing system that works for all products.

As for establishing a production control system, such as kanban, there are several methods you can use. One method is to collect ordered kanban at a regular interval (such as daily), substitute ordered kanban with the time required for in-house production (such as one day), and then issue in-house kanban with production specifications sent to the final process in the in-house production system. This final process becomes the base from which kanban are circulated back through whatever previous (upstream) processes exist in the in-house production system.

This method allows the supplier to respond to daily fluctuation in orders and to provide a period for delivery of products to buyers by maintaining a product inventory, and it also enables the supplier to

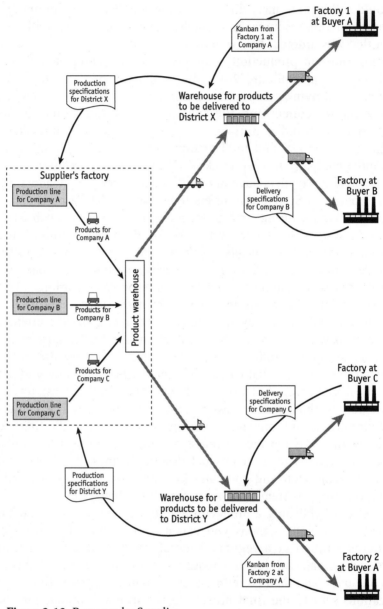

Figure 2-13. Response by Supplier

supply all ordered items, without any missing items. In keeping with the principles of the kanban system, each upstream process in the in-house production system will maintain its own store of in-process inventory.

Another method is to establish an in-house production schedule based on preliminary information from kanban that is issued by buyers three or four times a month. When using this method, the order information that arrives in the form of kanban or sequential delivery specifications, such as described in Figure 2-12, can only be used as delivery specification-related information. This method also helps to prevent missing items in shipments to buyers, and it usually involves maintaining an inventory of products or semifinished products.

Of course, no matter which method the supplier chooses to employ, the supplier must still bear the responsibility for accommodating the fluctuation in production variety and volume and for maintaining a sufficient stock of products and in-process inventory. However, when a supplier uses its own kanban system in correspondence with kanban systems of its buyers (the ordering companies), the supplier must consider only the information from the ordering companies as the correct information. This leaves very little wiggle room for in-house control, which forces the supplier to engage in a sort of seat-of-the-pants spot-market style of production control. This approach involves several risks, such as products being retained for a long time or becoming dead inventory, or the supplier needing large sums for equipment investment or allocation of labor.

THE ADVANTAGES OF DELIVERING TO MULTIPLE BUYERS OR FACTORIES

We have already examined the fundamentally conflicting rationales used by buyers and suppliers. This collision of rationales tends to occur most often when the supplier provides products to several buyers or factories, which creates problems and other challenges for production management.

When a supplier provides its own products to just one buyer, the supplier may be obliged to implement a thorough-going JIT production system with dedicated production lines for that final company. In such cases, no major problems or other difficulties are likely to occur in the supplier's production system as long as information about the buyer's ordered product types, product numbers, amounts, and so on, is communicated accurately to the supplier.

If the supplier delivers its products only to one buyer's designated production line at a designated factory, it can set up dedicated production processes that are fully synchronized with those of the buyer's production line. This means that the dedicated production processes will function as upstream processes for the buyer's production line. Here, the thorough implementation of the JIT Production System is the key, and should be considered a must for the supplier. This approach will effectively prevent any production problems that would be due to conflicting rationales between the supplier and buyer.

In addition, when a supplier is basically dedicated to a single buyer in this way, the buyer holds life-and-death power over the supplier, which means that the supplier is desperately obliged to serve the buyer's every need. To the extent that the supplier is successful in doing this, it will be spared problems in production management, and even in overall company management. However, only a handful of supplier companies provide products to just one buyer. The reality is that managers of suppliers would rather be able to seek their own profitability independently, which means they would rather not be wholly dependent on a single buyer. This way they can cultivate such profitable avenues as cutting-edge R&D in their specialty. Likewise, buyers do not usually depend upon one supplier. We'll explore this phenomenon further in the next two sections.

Suppliers Seek Multiple Buyers—Buyers Use Multiple Suppliers

As is well known, many of Japan's leading buyers have created a *keiretsu* structure, but even when such a keiretsu exists, suppliers that deliver products to only one buyer and depend on that arrangement as their entire business are still exceedingly rare. (*Note*: keiretstu is a group of Japanese firms that form historic associations and equity interlocks such that each firm maintains its operational independence but establishes permanent relations with other firms in its group. Some keiretsus are horizontal, involving firms in different industries, others are vertical, involving firms up and downstream from a firm that is usually a final assembler.) The main reason for this is that most of the buyers in Japan have adopted a procurement policy of purchasing from multiple sources. In other words, they practice multisource procurement, which means purchasing similar products (or parts/materials) from two or more suppliers. The purpose of this procurement policy is to keep the costs of purchased goods lower by encouraging competition among suppliers. It also helps to keep the relationship between the procurement staff and the supplier's staff from getting too cozy and complacent. Of course, there has been some news lately about how some companies have been making rapid progress in reducing the number of parts and materials suppliers as part of a cost-reduction campaign. But even in these cases they are not reducing the number of suppliers to just one company. Typically, the target number of suppliers is two or three.

Ordinarily, buyers practice multisource procurement at the level of similar parts (or part names), such as when an automobile manufacturer purchases the windshield glass for a certain car model or a combination meter for another particular car model. At this subordinate level, each part or material has a part number. For instance, the windshield glass for a certain car model may involve several different parts that each have their own part number, such as when the different parts (windshield glass types) correspond to different grades within a car model, each part may have different features, such as strength ratings, UV protection, or color tinting. The same holds true for other types of parts, such as combination meters, which vary depending on whether or not they include a tachometer, oil pressure indicator, turbo meter, and so on.

A buyer typically practices multisource procurement at this level of similar parts (or part names). This means that an auto manufacturer, as a buyer, may purchase windshield glass for a certain car model from Companies A, B, and C. The part numbers are used to allocate different part orders to different suppliers so that, for example, the buyer may be purchasing only high-grade UV-protected windshield glass from supplier Company A and only regular windshield glass from supplier Company B. This allocation of different parts from different suppliers tends to remain fixed until a car model change occurs.

For the suppliers, profitability may depend a lot on which part numbers the company can manage to provide to their buyers, since order volumes and profit margins vary among the different types and grades of parts. Since the buyer's (in this example, the automaker) production capacity is predetermined in terms of upper and lower limits, it must watch recent sales trends to determine which car models and grades are most likely to sell. Thus the orders (based on part numbers) that are received by suppliers can vary greatly from month to month. Also, even between model changes, the amount of parts that are actually ordered may be quite different from what was originally estimated.

Consequently, under this system based on multisource procurement, suppliers that provide products only to certain buyers may set mid- or long-term plans for their businesses, but securing profits in line with such subjectively set plans can be very difficult. Given the tough economic climate in which today's industries find themselves, suppliers are more challenged than ever to exercise independent management and secure profits on their own, even suppliers that belong to a keiretsu or have close ties to a certain division of a buyer company.

Consider the world's two largest U.S. automotive parts manufacturers: Delphi (No. 1), which became independent from General Motors, and Visteon (No. 2), which became independent from Ford. Both companies expanded their success by weaning themselves away from being dependent on just one company. These examples demonstrate how a supplier can benefit from weaning their business away from depending on supplying a single buyer. In other words, one of the best activities a supplier can pursue in the long run is developing multiple buyers.

Supplier's Strategy of Providing R&D

Another reason why few suppliers provide products to just one company is that suppliers tend to focus their engineering and product development resources on their particular specialties. As such, these supplier companies try to identify and serve a maximum number of target buyers for their products, not only to boost sales, but also to enable further cutting-edge R&D in their areas of specialization.

The best suppliers work hard to make the most of their engineering and product development skills in order to move into new market niches and gain more buyers. Managers at suppliers who have helped their companies cultivate these types of skills are adamantly against being simply suppliers to a particular buyer. Nevertheless, even the leading suppliers share a rather common weakness: while they may excel at R&D and new business development, such companies usually take only a superficial interest in the area of production management and, consequently, their level of production management technology tends to be low. As a result, their Production Division is regarded as relatively unimportant and the production managers are simply told to make and deliver products as the orders come in.

Challenges of Receiving Orders from Multiple Buyers or Various Factories

When a supplier receives orders from multiple buyers or various factories, it must still meet all of the buyers' orders without any delays or missing items, as well as deliver reliable, acceptable quality product. However, the information that such suppliers receive from their buyers and factories is typically based on varied order timing and ordering intervals, such as hourly, daily, weekly, or monthly orders. Furthermore, the accuracy of estimates for future orders also varies, including estimates based on confidential, in-house information, that the supplier receives from their buyers and factories. So suppliers have to interpret the estimates accordingly. At such times, head-on collisions can occur between the rationales of the ordering companies and supplier companies.

When faced with circumstances such as these, the people in the supplier's Production Management Division who are in charge of production planning tend to use the arguments described below as they seek to devise safe production schedules for the Production Division to follow (see Figure 2-14.) As reasonable as these arguments may appear at first glance, there should be no doubt that they are risky propositions indeed.

In today's era of wide-variety, small-lot production and increasingly short product life cycles, there is no guarantee anywhere that products left over from overproduction will be sold. Also, the

---

Arguments (Excuses) of Production Planners

1. "If you don't have the product in stock when an order comes in for it, you won't hear from that customer again."

2. "We have to maintain a lot of product inventory in our warehouses, or we won't be able to handle the wide array of orders we get from our customers."

3. "We have a lot of inventory, and I know that's not a good thing. But our products aren't perishable—whatever doesn't sell right away we can sell later on. Consequently, we can adjust our inventory levels according to how sales are going so that eventually we will use up all of our inventory."

4. "Making products in large numbers helps us reduce costs through economy of scale, which helps us maintain low sales prices and remain competitive in the marketplace.

Figure 2-14. Arguments (Excuses) of Production Planners

principle of reducing costs via the economy-of-scale effect works only when you have sold virtually all of the mass-produced products. Having lots of remaindered products on hand not only hurts the company's cash flow, but it can also deal a critical blow to the company's bottom line.

Recognizing the pitfalls of such safe thinking, managers at suppliers have turned toward the JIT Production System as a promising new strategy. As a result, Production Management Division managers have narrowed their management focus to the point where their main management function is to simply pass along information from the received orders to the Production Division. Meanwhile, in the Production Division, production orders are issued based on this information, which means that production orders fluctuate as much as the buyers' orders do, and this fluctuation is directly translated into daily ups and downs in the production yield, labor hours, and other shopfloor conditions. The workload for the Production Division staff also increases greatly, since they are the ones who must order and manage the parts and materials needed to meet the buyer' orders (see Figure 2-15.)

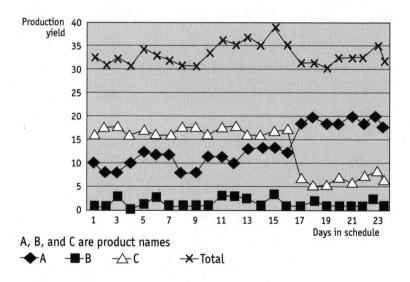

A, B, and C are product names
◆A   ■B   △C   ✕Total

Figure 2-15. Example of Production Schedule Based Strictly on Buyers' Orders

When dealing with a production schedule such as the one shown in Figure 2-15, Production Division managers work in terms of the 4Ms (man, machine, material, and method) by paying for overtime labor when the production load is heavy and paying for underutilized resources when the production load is light. An increasing number of companies have found that keeping up a schedule like this requires extra work on the part of the parts/materials managers and transport/conveyance experts, so that even if improvements are made in the way that equipment operators do their work, the Production Division's labor productivity is still likely to decline.

Incidentally, the results of comparative studies of orders from various buyers have shown that there is less fluctuation in orders issued by buyers who carry out planned, schedule-based production. Fluctuation tends to be greater in orders issued by buyers who have introduced the JIT Production System and are consequently issuing kanban orders. These results are yet another indication of how suppliers are being exploited by the kanban system.

It should be added that orders from buyers who have correctly and appropriately introduced JIT are more stable than the widely fluctuating orders from buyers who have formally introduced JIT but are not managing their kanban system correctly. The causal factors and current problems related to poorly implemented kanban systems are described in the next section.

HOW BUYERS' INCORRECT USE OF KANBAN EXPLOITS SUPPLIERS

The kanban system is a technique used when implementing the JIT Production System whereby a manufacturer produces just what is needed, just when it is needed, and just in the needed amount. Figure 2-16 describes a typical example of the problems faced by a supplier that is being exploited by a buyer's kanban system. The problems are even more severe when the buyer implements kanban incorrectly. Still, the problems that suppliers face when their buyers are implementing the kanban system correctly differ only in degree from the problems they face when their buyers are implementing the kanban system incorrectly.

As mentioned earlier, when a buyer operates a kanban system in an incorrect or inappropriate way, it is known as an imitation kanban system. How do imitation kanban systems occur? The roots of the correct JIT Production System lie in the principles of production leveling and autonomation, or automation with a human touch (i.e., automatic control of defects). An imitation kanban system tends to ignore these two basic principles (see Figure 2-17.) Both of these

1. The buyer, which has more leverage in the relationship, demands immediate delivery response from the supplier company based solely on its own needs, and the supplier must comply.

Poor application of kanban rules

2. When a supplier has among its customers several buyers that have introduced a kanban system, it must ignore production efficiency and consequently put up with large amounts of waste in its production system in order to respond to the various kanban issued by those buyers. Wastes would include pure waste, waste due to unevenness, and waste due to unreasonableness.

Deviation from the purpose of the JIT Production System

3. The supplier must maintain a certain amount of "safe inventory" so that it can respond promptly to changes in order amounts and delivery deadlines (timing). The supplier tends to make the inventory excessive due to uneasiness about the ability to respond promptly.

Deviation from the purpose of the JIT Production System

Figure 2-16. Example of Problems Faced by a Supplier Being Exploited by Kanban

basic principles correspond to the Plan stage of the PDCA (Plan-Do-Check-Act) management cycle.

Production leveling means working at the production planning (production management) stage to devise a production schedule that enables the most efficient use of production resources such as the 4Ms (man, machine, material, method). *Autonomation* means devising ways to make full use of production resources so as to keep fully in pace with the production schedule during the production implementation stage. An example of this wise use of production resources is to stop production, or to have production automatically halted once a certain output has been reached to avoid producing more than was indicated in the production specifications. Another example is to stop production, or have it automatically halted whenever a defective product is detected, so that you can only produce nondefective products. The preliminary studies and arrangements that must be made at the production management stage and the production implementation stage both correspond to the *Plan* stage of the PDCA management cycle.

By contrast, an imitation kanban system focuses only on the *Do* stage by making production schedules easier to plan and making

How do imitation kanban systems occur?

• Imitation kanban systems only look like true kanban systems

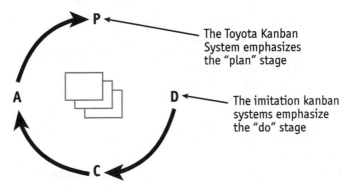

The Toyota Kanban System emphasizes the "plan" stage

The imitation kanban systems emphasize the "do" stage

Imitation kanban system/Toyota Kanban System

Figure 2-17. The Main Difference Between a True Kanban and an Imitation Kanban System

parts and materials easier to order. Such a system is operated with a continual concern to find similar labor-saving shortcuts. Figure 2-18 digs a little deeper into the causes of imitation kanban systems by listing three areas in which companies use shortcuts. We'll now discuss each of these areas briefly, including some background factors.

Shortcuts at the Production Planning Stage

• **Precision of rough estimate orders.** First, under the banner of being consumer (market) oriented, the buyer's Production Management Division staff recognizes revisions in their production schedules as obviously necessary, so they usually create a low-precision (i.e., rough) production schedule without putting adequate preliminary research into it. At such companies, the production managers figure that they will need to adjust the production schedule later on anyway, in response to actual demand from the consumer and production conditions.

Secondly, when the buyer's Production Management Division presents the various manufacturing departments with a basic production schedule, they make the Manufacturing Department staff responsible for finalizing the actual production schedule and ordering the required parts and materials.

When a buyer takes these kinds of shortcuts at the production planning stage, its production planners consider it only natural that the production schedule and order specifications will need to

---

1. Shortcuts at the production planning stage:
   • Precision of rough estimate of orders

2. Shortcuts in kanban system maintenance:
   • Setting of initial standards
   • Maintenance (revision) of values in standards due to equipment modification or use of new equipment
   • Maintenance of bills of materials
   • Maintenance of total required amounts

3. Shortcuts in operation of kanban system:
   • Corrective action in response to abnormalities

Figure 2-18. Three Shortcuts that Are Typical of Imitation Kanban Systems

be revised later on, and the supplier companies are well aware of the likelihood of such revisions. Or, they figure that since the actual, definite orders will be provided via the kanban system (by withdrawing kanban for each order), the required parts and materials simply will be ordered and supplied without anyone having to discuss it.

Shortcuts in Kanban System Maintenance

- **Setting of initial standards.** Often the buyer's Production Management Division does not really understand kanban. They think it is enough just to use kanban according to their supervisors' instructions when ordering products, parts, or materials. At a company such as this, kanban has been introduced only in form, but some of the substance has been omitted due to the shortcuts that were taken when setting initial standards for the kanban system (see Figure 2-19).
- **Maintenance (revision) of values in standards due to equipment modification or use of new equipment.** The buyer's Production Management Division must maintain various standards in a timely manner when changes in products or processes occur, such as when existing production equipment is modified or new production equipment is installed (see Figure 2-19). Since maintenance of values in standards is a tedious, time-consuming task, it is an area where shortcuts are likely to be taken.
- **Maintenance of bills of material.** As design changes are made and new products are introduced, the buyer's Production Management Division must make corresponding changes in their lists of required parts and materials, which includes maintenance of

The following items should be verified specifically for each part or material. The number of kanban in circulation should be set specifically for each part or material after these initial standards have been set.
• Kanban cycle
• Safe inventory levels
• Per-day consumption
• Capacity

Figure 2-19. Example of Initial Standards for Kanban System

bills of material. Since this also a tedious, time-consuming task, it is an area where companies are likely to take shortcuts.

• **Maintenance of total required amounts.** The buyer's Production Management Division sometimes doesn't even recognize the need for this type of maintenance, since it believes the kanban system precludes the need for deployment of parts (creating parts lists) and calculating total required amounts. The fact is that, even when you are using a kanban system it requires both parts deployment and the calculating of total required amounts. It is only the day-to-day, ad hoc deployment of parts that is alleviated by use of the kanban system. Consequently, to use the kanban system correctly, you must deploy parts based on the parent part numbers for the product in question, and then estimate the total amounts. Also, you must check up on the process to make sure the estimated total numbers of parts and materials are not too much or too little. This would include things such as the confirmation of number of kanban in circulation. Because companies misunderstand total required amounts, or are unwilling to devote labor to it, they are likely to take shortcuts in this area.

Shortcuts in the Operation of Kanban

• **Corrective action in response to abnormalities.** If the buyer's Production Management Division and Manufacturing Division are using their kanban system correctly, it should make it easy for anyone to tell that there are missing parts or materials or excess inventory when a problem occurs. If it is hard to tell when there are missing parts or materials or excess inventory, or if only people who are directly involved can identify and correct the problem, the company is not using its kanban system correctly. To put it another way, the kanban system provides ways to make the occurrence of abnormalities obvious enough to serve as a signal for a needed improvement. The essence of the kanban system is the detection, thorough causal investigation, and remediation of abnormalities through improvements that eliminate their causes and thereby help prevent their recurrence.

From the perspective of production management, it is a bad indicator if the occurrence of missing parts or materials or excess inventory occurs after an abnormality. Unless the managers understand what the kanban system is really about and operate it accordingly, shortcuts are likely to be taken. For example, it would be a shortcut if, when faced with a problem indicated by missing parts or materials (according to the number of circulating kanban), the buyer's managers, without even first investigating the cause of the problem, attempted to fix it by calling or faxing the supplier to explain that they should make a delivery without a corresponding kanban. Also, a prudent buyer would have spare kanban or extra printed kanban to use in such cases instead of no kanban at all. Figure 2-20 lists some typical causes of missing parts and materials (i.e., a mismatch with the number of circulating kanban).

---

1. Lost (misplaced) kanban.
2. Error in setting or updating standard values for kanban.
3. Error in bill of material.
4. Error in total required amount.
5. Defective parts or materials.
6. Occurance of defects at final company.
7. Supplier's delivery error.
8. Inadequate enforcement of rules for kanban operations:
   (*Example 1:* Kanban are held back at the supplier)
   (*Example 2:* Kanban are held back at the buyer)

Figure 2-20. Causes of Mismatches with Numbers of Circulating Kanban

SHOULD YOU USE THE PUSH METHOD OR THE PULL METHOD?

The original form of production control was a set of three functions—planning, specifications, and regulation—that were all the responsibility of the manufacturing company in question. The basic production control functions outlined in Figure 2-21 do not change, regardless of whether the production system is JIT, ESP, or any other production system. As the figure shows, production control includes ideas that are decided upon, implemented, and evaluated, all by the manufacturer. In other words, the manufacturer determines its own production control methods, then implements and regulates them.

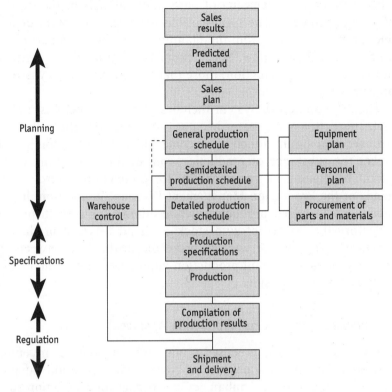

Figure 2-21. Basic Production Control Functions

This means that the manufacturer bears the responsibility for the entire process, and the manufacturer's *subject* at every stage from decision-making to evaluation is *itself*. From this perspective, if one were to ask "Which is better, the push production approach or the pull production approach?" the answer would be push production. However, the reader should understand that this question, which compares the push production approach and the pull production approach is actually a meaningless question in this context. This is because whether the production approach is push or pull, the basic production control functions remain the same. Nevertheless, if one insists on choosing between one approach or the other, the push production approach would be the better choice, since it fits more deeply with the substance of the production system.

The distinction between the push production approach and the pull production approach is simply a distinction between two ways of administering production control. They differ mainly in terms of what triggers the purchasing of parts and materials or the starting of a production run. The specifications for a certain factory or production process come from the planning division, or as part of the processing/assembly flow. Specifications that come from upstream processes = push production; those that come from the downstream process = pull production.

The JIT Production System includes planning functions and, in fact, the better a company becomes at using JIT, the fuller and stronger its planning functions become. The ESP Production System makes the substance of production control functions more transparent by placing all responsibility for production control in the hands of the Production Management Division. In addition, it promotes synchronized production while aiming for zero in-process inventory (i.e., stores of zero stock on hand between production processes). Thus, if we were to make the distinction between the push production and pull production approaches, ESP would be on the push production side.

The Kanban System—A Pull Production Approach

The kanban system is premised upon having in-process inventory (stores). It falls into the production control administration category of pull production. The kanban act as a trigger for production specifications issued in a pull production system. When an item is pulled by a downstream process, a kanban corresponding to that item in the in-process inventory store is detached, at which point it serves as a trigger for starting production at the process in question (see Figure 2-22).

Thus, under a kanban system, when a part is used at a subsequent (downstream) process, a withdrawal kanban is detached so that the same part can be withdrawn from the store of in-process inventory at the previous process. When the part is withdrawn from this inventory at the previous process, another withdrawal kanban is sent to the process before that one as a type of production specification. The basic principle of the kanban system is to link processes in this way while production is being carried out.

In view of this basic principle and the illustration shown in Figure 2-22, the kanban system is, in the final analysis, a system for

Production specifications

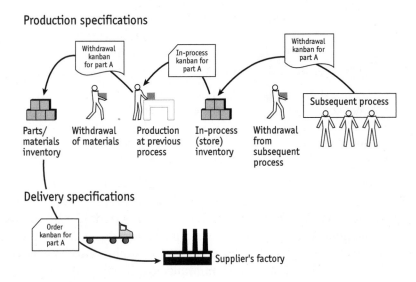

Figure 2-22. Pulling of Kanban and Production Specifications

replacing parts that have been used. As such, it is a system designed to use product inventory effectively and it therefore requires some level of in-process inventory, although there is an effort to keep that level as low as possible.

Several techniques are used as part of this effort to minimize in-process (store) inventory, including the determination and enforcement of production specification rules for the sake of production leveling at downstream processes, detecting and correcting defects at each process, improving changeover (retooling) procedures at each process, reducing lot size, and minimizing capacity differences between processes. In any case, the kanban system is a system designed to use inventory effectively. By definition, it can never achieve zero inventory.

The true purpose of the original kanban system

Kanban is basically a set of specific tools that are used under the JIT Production System. As mentioned earlier, other components of JIT include production leveling and autonomation. One of the tricks used in autonomation is to stop production at an individual process once that process has produced a certain amount of yield, in order to prevent overproduction. The full work system is one of the recent variations on this theme.

However, the most important function and true purpose of autonomation is to identify abnormalities that occur and are indicative of defects, and to signal that you need to make an improvement. In other words, the essence of kanban is the process whereby:

- Abnormalities are immediately detected when they occur.
- Production is stopped right there and then.
- The causes of the abnormality are thoroughly investigated at once.
- Improvements are made to eliminate those causes.
- Measures are taken to prevent the problem from occurring again.

From the perspective of production control, though, a trick that stops production whenever an abnormality occurs is not a very good trick. What is the point of stopping production, identifying the phenomenon, and signaling the need for an improvement whenever an abnormality occurs? For example, if the abnormality is a defect, is all of this being done just to minimize the number of defects? To be sure, there is more to it than this shortsighted purpose.

The greater purpose can be seen in the wisdom of the Toyota Production System. As is widely known, production lines in automotive assembly plants tend to be very long, very numerous, with many processes in each line. There are also various groupings of main lines and sublines (the latter would include lines for assembly of engines or instrument panels), along with stand-alone processes (such as press processes), all arranged in a complicated layout. Obviously, such plants contain large numbers of machines and employees. In plants such as these, it is vital to be constantly on the lookout for abnormalities that may occur in any process, in any piece of equipment, or with any employee, and to maintain control to prevent these abnormalities.

That is why production managers at Toyota encourage all plant employees to report abnormalities. Specifically, they are told to stop production and raise their hand whenever an abnormality becomes apparent, with encouraging words such as abnormalities are not necessarily a bad thing. We actually want to detect abnormalities so we can prevent their recurrence. Then the manager will add that the recurrence of abnormalities is something I do not tolerate. When they occur again, I get mad. Toyota has thus used this approach to thoroughly improve its production processes, thereby creating one of the world's leading production systems.

What the reader should understand better than anything else is that kanban is more than a tool used for production control—it is an indispensable technique that was developed within the environment of an automobile manufacturer, a particular type of manufacturing company. So, without fear of being misunderstood, let it be said that the true purpose of the kanban system is to implement improvements promptly and reliably and to promote ongoing improvement.

ERP, SCM, AND THE ESP PRODUCTION SYSTEM

ESP serves a key role in building reliable Enterprise Resource Planning (ERP) and in developing ERP-backed Supply Chain Management (SCM). Generally, the ERP and SCM systems that are currently in use have been judged as effective only for production systems that involve a narrow or weak range of production control methods. In particular, the ERP system, which grew out of an accounting approach, has been criticized for various problems it poses as a production control technique and for being difficult to use.

However, when you introduce ERP or SCM it should be used as a package as long as there are no firmly established production control concepts or practices that would preclude such package usage. In fact, the package-introduction approach is strongly demanded as part of an ERP or SCM system. As a result, companies have been obliged to implement and run a ready-made production control system, even though this has led to some serious practical difficulties in many cases.

The ESP Production System offers an alternative that can prevent such difficulties. When a company implements and deploys ESP, it puts in place production control techniques that fit the company's own needs. This is because of ESP's relatively simple approach, whereby the central leadership role of the Production Management Division, and its control and administration methods, make comprehensive use of patterns and quantities. It also basically eliminates all constraints on production processes. Consequently, ESP naturally lends itself to computerization, and can act as a reliable supplement to techniques such as ERP and SCM. Figure 2-23 lists four key points concerning the implementation of ERP, all of which are compatible with ESP.

In Chapter 1, we introduced the original impetus behind developing ESP and some of the early concepts. In this chapter, we expanded on ESP's roots, with its Four Principles, and explained why suppliers that are ill-suited for JIT or deal with multiple buyers, need it to

- ERP is a switch in production control from "people-centered, follow-the-leader control" to more integrated and transparent "system-centered management."
- To make this switch to "system-centered management," all existing work processes and procedures must be fundamentally reevaluated and rebuilt.
- The resulting system must be simple.
- If possible, the rebuilt system should be applied not just domestically but globally.

Figure 2-23. Key Points Concerning the Implementation of ERP

survive. Now you are ready to dig a little deeper and look at the basic steps for implementing ESP, which entail learning about the Six Guarantees—the concepts behind achieving the objectives of ESP. In fact, in the next chapter you will see why maintaining and expanding upon all six of these guarantees are essential for successfully implementing ESP.

3

Basic Steps in Implementing the ESP Production System

The concepts behind achieving the objectives of the ESP Production System are summed up as the Six Guarantees (see Figure 3-1). In fact, the main objective of ESP is to implement and maintain the Six Guarantees while continually devising ways to further expand them. Of course, this ongoing maintenance includes responding to changes, such as changes in the business environment or in business practices, in the list of buyers (added/dropped buyers, etc.), in

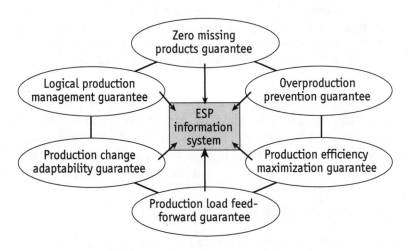

Figure 3-1. The Six Guarantees of the ESP Production System

the types and amounts of products being ordered, and in new product lines, as well as new designs for existing products. In addition, the company must respond promptly and reliably to any innovations or improvements of its own manufacturing processes. It is also important that the Production Management Division anticipate such innovations and take the initiative in responding to them. Only when the company adopts this kind of proactive stance toward change and improvement can you say it is continually devising ways to further expand the success of its production system.

To achieve this proactive stance toward change and improvement, ESP devised the Six Guarantees. In fact, all six of these guarantees are absolutely essential for fulfilling any manufacturing company's production management functions. In other words, ESP is premised on following the sequence of the Six Guarantees so that a supplier can:

- Guarantee zero missing products (including zero late deliveries) as the primary evidence of its reliability, *while*
- Working to maximize production efficiency, *by*
- Effectively using limited inventory (controlled minimum inventory), *by*
- Using equalized units for product item numbers in small lots, *by*
- Carrying out synchronized and equalized production, *which*
- Helps to minimize variation in the production schedule's daily production yield, *and*
- Enables flexible responses to rush orders and canceled orders, which includes using computers to achieve these objectives, *while*
- Establishing production management logic early on, in coordination with the relevant manufacturing companies, *thereby*
- Weaning the company away from a production management system that is dependent upon an inner circle of managers.

It is a sure bet that any manufacturing company that manages to ensure the Six Guarantees—all essential ingredients for ESP's success—will gain the rock-solid trust of its buyers and will be able to depend on that trust as the company continually strives to improve. We will now describe the Six Guarantees in further detail as they pertain to the supplier-buyer relationship, as discussed in Chapter 2.

GUARANTEE ONE—ZERO MISSING PRODUCTS

The first and foremost ways for suppliers to gain and maintain the trust of their buyers is to make sure there are no missing products.

Simply put, it is the supplier's job to respond to buyers' orders by delivering the ordered products, in the ordered amounts, on the agreed-upon delivery date. A reputation for delivering shipments with missing products can deal a fatal blow to a supplier. That is why the Zero Missing Products Guarantee is listed first and should be first in the mind of your production managers when setting out to implement the ESP Production System.

Under ESP, a supplier can uphold its Zero Missing Products Guarantee by treating the preliminary in-house order information or the confirmed order information provided by the buyer as order information within previously established ESP production patterns. (As you may recall, in Chapter 1 we defined the ESP production pattern as a kind of early, preliminary production plan that helps to clarify the supplier's production planning standards and anticipates demand from buyers. Or another way of looking at it, is as a type of substitution formula used to analyze scenarios based on proposed manufacturing of goods. This will be discussed in greater detail later in this chapter. The supplier now performs a computer simulation of the production, shipment, and inventory levels before sending production specifications to its factories, which enables it to establish a production schedule designed to avoid missing products. In this way, the supplier can keep tabs on its production capacity and pending orders, while continually reviewing and updating its production schedule. As long as the factories can manufacture products according to this ESP production schedule (i.e., a confirmed production schedule), the supplier can rest assured that there will be no missing products. The ESP production schedule, which utilizes ESP production patterns, is the basic principle of production scheduling.

In addition, by maintaining limited inventory (described later), the supplier can avoid having missing products even after production based on the buyer's order information has been started. For example, in situations where changes in the buyer's order information would otherwise make it impossible to produce all of the ordered products within the agreed-upon delivery lead time, or when the buyer calls with order changes even after the ESP production schedule has been sent to the factories.

GUARANTEE TWO—OVERPRODUCTION PREVENTION

For a supplier, a lot of excess inventory can threaten its survival. As was described in Chapter 2, retaining goods for a long time is not only bad for the supplier's cash flow, it entails an incalculable risk that any or all of the retained inventory may end up becoming dead inventory. Furthermore, when a supplier has an excess inventory of products, in-process inventory, parts, and materials, it must expend considerable time and money just to manage the excess inventory. The Overproduction Prevention Guarantee is an essential part of the ESP Production System because it provides a procedure to prevent excess or otherwise unnecessary inventory.

Under ESP, after a supplier performs the simulation under the first guarantee, it runs a second simulation. These results are then used to draft a new production schedule. Now the supplier is in a position to prevent overproduction by using the simulation results to stop production (described later) whenever the results indicate that a product's limited inventory level is going to be exceeded.

GUARANTEE THREE—PRODUCTION LOAD FEED-FORWARD

Load leveling is indispensable as a means of steadily boosting and maintaining the efficiency of manufacturing processes. Leveling (Heijunka) eliminates unevenness in production loading (combinations of product variety and quantity) of each production order and the timing of the production order. In short, "load leveling" is to periodically issue a production loading schedule that has no unevenness in both product variety and production amount. As long as the production load is allowed to fluctuate from day to day at various manufacturing processes, it becomes unlikely that the supplier can make effective use of its limited production resources, particularly labor and equipment resources. Using ESP's Production Load Feed-Forward Guarantee will provide the supplier with a straightforward and rational load leveling technique to help it level its production load.

Generally, the first step in leveling the production load is to analyze the load variation that has occurred as the load has accumulated. Then you do the leveling by feeding some of the load either forward or backward. However, under ESP, the basic load leveling method is to feed the load forward, since this method works best for simultaneously accomplishing the Zero Missing Products Guarantee. Before a supplier can implement ESP's Production Load Feed-Forward guarantee, the supplier must first perform the simulations for the zero missing products guarantee and the overproduction prevention guarantee. Only then will it be ready to perform a third simulation to determine whether or not the production load is under or over the current production capacity (on a scale of 0 percent to 100 percent).

If the results of this third simulation indicate any areas where the production load exceeds 100 percent of the production capacity, the load is leveled by being fed forward according to the ESP production patterns while remaining within the confirmed period of the ESP production schedule. If this feed-forward adjustment is not enough to reduce the production load to the level of the production capacity, you must take exceptional measures, such as assigning overtime production to handle the remaining overload.

As for areas where the production load is substantially less than 100 percent of the production capacity, you will use ESP production patterns to determine how to feed some products (item numbers) forward without causing production problems to level out the production load, while remaining within the confirmed period of the ESP production schedule.

Under ESP, the load is adjusted to create a production schedule that incorporates a feed-forward production load. As a result, the ESP production schedule, that is, the confirmed schedule that is sent to the Manufacturing Division, is a leveled production schedule, without any significant load variation within the scheduled period. To achieve this kind of feeding forward of the production load, production capacity must be either equal to or greater than the production load (the amount of orders). Naturally, the supplier must review and update the production capacity indicated in the ESP production patterns at a regular interval (such as quarterly or semi-annually) or whenever a buyer changes its average level of orders.

If it becomes apparent when reviewing ESP production patterns that the load (amount of orders) is increasing, then the supplier must consider taking various measures to correct it. This would include such things as making improvements to expand the production capacity or changing the operation hours (revising the regular shifts or setting up overtime shifts), after which the supplier would accordingly revise the production schedule and the corresponding ESP production patterns. Conversely, if it appears that the load is shrinking, the supplier is then obliged to consider measures such as consolidating certain production processes or shortening operation hours, after which it must once again revise the production schedule and ESP production patterns accordingly.

At many suppliers, the approach to production leveling is to estimate the load based on the delivery dates that have been set for each buyer when accepting their orders, after which the production managers adjust the load to make it as level as possible. For this reason, most suppliers rely a lot on the skill and experience of their production managers to perform this kind of load leveling work. Which means, more often than not that no rational load leveling technique is ever spelled out. Needless to say, the production managers spend a lot of time figuring out how they can level out the production load, and for many it is a major headache. To eliminate the guesswork, the ESP Production System provides the following policies, which we will describe later, concerning production load leveling.

- **Limited inventory concept** enables buyer orders to be considered separately from the production schedule.
- **ESP production patterns**, when set, are used in tandem with orders received from buyers to perform various simulations before drafting the ESP production schedule.
- **Production sequences and equalized units** are included in the ESP production patterns to facilitate the three simulations described in the first three guarantees.

By establishing rational load adjustment methods, these policies enable the work of leveling the production load to cease being something that only an inner circle of experienced production managers can perform.

GUARANTEE FOUR—PRODUCTION EFFICIENCY MAXIMIZATION

Maximizing production efficiency is a never-ending issue for manufacturing companies. When production efficiency is at its highest, it not only reduces manufacturing costs, but also shortens production lead time and helps to cut down on inventory levels of products, in-process inventory, parts, and materials, all of which work to boost the company's profitability and improve its cash flow. As discussed in Chapter 2, foremost among the Four Principles behind ESP is Principle One, according to which everything is decided at the planning stage. And recognizing that productivity (i.e., the basic level of productivity) is mainly determined at the production planning stage means that the way you establish the production schedule determines the degree of production efficiency, thereby optimizing production efficiency at the production planning stage. The Production Efficiency Maximization Guarantee is the concrete expression of this principle and will help you secure these outcomes. In other words, since the basic level of production efficiency is thought to be determined by the production amount (load) indicated in the production schedule, one of the main points in implementing the Production Efficiency Maximization Guarantee is to establish and maintain a production schedule that uses 100 percent of the production capacity.

Under ESP, the supplier issues production specifications after setting up a production sequence that is oriented toward high production efficiency and proposing an ESP production schedule that incorporates this production sequence. The Manufacturing Division then must comply with the ESP production schedule and run the production lines accordingly. As was mentioned in Chapter 2, the reason why you must maximize the production efficiency at the production planning stage is because getting a detailed grasp of the Manufacturing Division's production conditions at that stage will enable planners to draft a production schedule that effectively minimizes line stoppage loss, velocity loss, and defect loss.

The first thing that you must consider is model changeovers, such as die replacements, color changes, and other changeover or adjustment tasks that may pose problems in wide-variety, small-lot production lines. Taking machine tool processes as an example, the production sequence is set so as to minimize the amount of time spent on changeover and adjustment work, such as switching between larger and smaller product sizes or replacing jigs and tools, and the production schedule is set according to the optimized production sequence. With regard to painting processes, the production schedule would include an optimized production sequence that goes from lighter colors to darker colors. And as for sheet rolling processes, it would include an optimized production sequence that goes from wider to narrower sheet sizes. For assembly processes, it would include an optimized production sequence that would help clarify which assembly parts are to be used so as to avoid assembling products with the wrong components.

In this way, the supplier establishes input sequences (production sequences) that are clearly based on the principle of maximizing the efficiency of its own manufacturing processes. The establishment of a production schedule that incorporates these kinds of optimized input sequences (production sequences) helps to keep various kinds of loss at a minimum, including stoppage loss related by changeovers, adjustments, and breakdowns; velocity loss related to the failure to maintain standard times; or defect loss related to quality problems and lower yields.

The supplier creates ESP production patterns that include input sequences (production sequences) based on the principle of optimizing its own manufacturing processes, and that also include equalized production units, (described later in this chapter) per product model (item number). As such, ESP production patterns are a fundamental element in the Production Efficiency Maximization Guarantee.

GUARANTEE FIVE—PRODUCTION CHANGE ADAPTABILITY

It frequently happens that there are differences between the preliminary in-house order information and the confirmed order information provided by the buyer, in terms of the indicated product types and amounts. Often, the buyer changes its order information right before the production input date. These types of changes are known as rush orders and canceled orders. Suppliers are simply expected to deal with these last-minute requests from buyers, no matter how inconvenient they are. To fully respond to such requests, even when a buyer requests changes in an order right before the production input date, ESP developed the Production Change Adaptability Guarantee. This essential guarantee establishes prompt and reliable revision functions (adaptable planning at the production planning stage) that a supplier can use right before the production specifications are issued in your ESP production schedule, which are based on the supplier's ESP production patterns.

How does a supplier implement this Production Change Adaptability Guarantee under the ESP Production System so that it can revise ESP production patterns at any time? First, as a policy under ESP, the supplier will take the equalized units that are based on production units (quantities) for each product item number and perform stops and insertions to revise (rearrange) ESP production patterns. As a result, the supplier can rationally revise its production schedule and implement it more quickly.

Using the Stop Method

1. The supplier will perform a stop if, after incorporating the buyer's order information into the relevant ESP production pattern and running a simulation, the simulation indicates certain item numbers will be overproduced.
2. The supplier will perform a stop if the simulation, based on the ESP production pattern and calculated product inventory (esti-

mated inventory), selects additional item numbers that need to be produced in response to a buyer's rush orders

Points concerning administration of stops

- You administer stops using equalized units, as determined by the production unit (quantity) for each item number.
- To feed-forward the load or to maintain a freely adaptable production schedule, the production scope of stopped item numbers can be item numbers that are needed to avoid missing items.
- If a buyer's orders tend to fluctuate regularly, you can set a certain production scope in advance, as for order fluctuation, to establish a nonspecific prearranged stop.

Using the Insertion Method

1. A supplier performs an insertion to add items if, after incorporating the buyer's newly inserted order information into the relevant ESP production pattern and performing a simulation, the simulation indicates certain item numbers will be missing.
2. A supplier performs an insertion to augment production as required by schedule revisions, that is, production in response to last-minute additional orders and rush orders.
3. A supplier performs an insertion to augment production when the limited inventory fails to meet the specified standard value (upper limit value).
4. A supplier performs an insertion to augment production of items needed for a feed-forward production load when an insertion scope exists, even though responses are being made for missing items or schedule revisions, or items are being returned to the limited inventory.
5. A supplier performs an insertion if a certain steady production scope has been set in advance as for order fluctuation (as a stop), to make up the difference within the production scope. (It can also be used to augment production in the situations described in 3 and 4 above, if the amount of order fluctuation is not being fully offset.)

Points concerning administration of insertions

- Insertions are administered using equalized units as determined by the production unit (quantity) for each item number.

- If the insertion scope is not being met, a suitable range of production can be fed forward at the current simulation stage so that you can make a periodic and centralized production stop.

GUARANTEE SIX—LOGICAL PRODUCTION MANAGEMENT

Unfortunately, it is not uncommon to find suppliers where the practical administration of production management functions (especially production planning functions) is completely dependent on the personal skills of individual members of the Production Management Division. This is one reason why so few suppliers use computers to administer or assist with the administration of production management tasks in general, and production planning tasks in particular. Under ESP, the Logical Production Management Guarantee simplifies production management functions so that a supplier can organize them logically instead of relying on the wisdom of an inner circle of experienced managers to carry them out. In other words, the Logical Production Management Guarantee enables suppliers to implement and logically manage production according to a rational set of production management criteria, allowing them to respond to actual conditions as they occur. The result is that the Production Management Division truly becomes more efficient, and it becomes possible to build a computer-based production system that can be further developed systematically.

BASIC STEPS IN IMPLEMENTING THE ESP PRODUCTION SYSTEM

We will now discuss the three basic steps for implementing ESP: 1) limited inventory (separating your buyer orders from your production schedule, 2) synchronized and equalized production, and 3) equalized amounts (an original concept in the ESP Production System). Once we introduce these, we will return to a discussion of the steps for realizing the Six Guarantees, and establishing your ESP production patterns.

Limited Inventory: Separate Your Buyer Orders from Your Production Schedule

As was mentioned earlier, ESP recognizes that suppliers cannot effectively control their production efficiency if they allow fluctuations in buyer orders to be directly reflected in their production processes, thereby creating peaks and valleys in production output. Therefore, ESP proposes that you maintain a limited inventory (limited amount of product inventory) to provide a buffer between buyer orders and the supplier's production schedule.

Having this limited inventory enables the supplier to immediately respond to buyer orders or market needs while preventing missing items in deliveries. Under such a production system, the manufacturing processes are not like vending machines: one cannot expect to just drop some money (or buyer orders) into them and immediately receive the desired products. The limited inventory is there to help the supplier deal with the unexpected. It also greatly helps the supplier to achieve what should be its primary goal in order to maintain trust and to survive—namely, the goal of zero missing items in deliveries.

As we have emphasized, the main objective under ESP is to balance the need to respond immediately to changes in buyer orders and market needs while establishing and maintaining production schedules that maximize production efficiency. The means used to reach this objective include maintaining a limited inventory (the minimum required product inventory), which is determined according to the supplier's manufacturing process conditions, such as manufacturing lead time and manufacturing process restrictions, as well as the amount of buyer orders and the fluctuation in buyer orders (see Figure 3-2.) Limited inventory is defined as a controlled minimum required product inventory. Basically, this limited inventory consists of finished products. Why does the limited inventory basically consist of finished products? Because having a lack of in-process inventory is a fundamental characteristic of the ESP Production System. ESP's policies for issuing production specifications and managing the progress of production without maintaining in-process inventory are described below.

Confirmed production schedule and in-process inventory

The confirmed production schedule, or ESP Production schedule, is specified for the first process or, if necessary, at other processes as well. Then production proceeds sequentially from this first process in a push production system, in which there is, in principle, no generation of in-process inventory. The ESP Production schedule is designed for maximum production efficiency, and the designated production division is generally responsible for creating it. This is

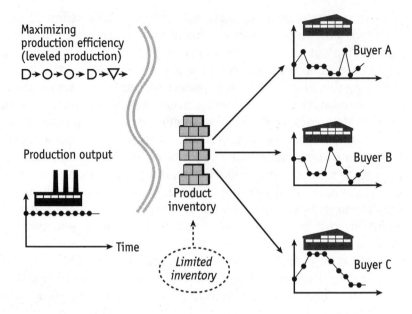

Figure 3-2. Limited Inventory: Separating Buyer Orders from the Production Schedule

how a supplier creates the production schedule so as not to generate in-process inventory other than that required for the specified products. The reader might wonder why the ESP Production System avoids having any in-process inventory.

One reason is that ESP requires that the Production Management Division maintain accurate accounting and control of all inventory. Otherwise the supplier will not be able to devise a production schedule that maximizes production efficiency. As a practical matter, if it were left up to the Production Division, rather than the Production Management Division, to maintain accurate accounting and control of all in-process inventory, it would be a difficult task requiring a very expensive set of control measures.

The second reason for having zero in-process inventory is in order to achieve the goal of making all inventory apparent. The concept of visible inventory is a basic principle of production management. It is difficult to get a solid, direct grasp of total inventory if in-process inventory is spread out among numerous production processes. By making it so that the only inventory is a store of finished products, the manufacturer can quickly and easily understand the current inventory levels. The factors that give rise to the occurrence of inventory are also easier to identify, since they can all be found in the inventory of finished products.

The third reason for having zero in-process inventory concerns the pursuit and building of a zero inventory production system. When a manufacturer has a limited inventory, the various reasons for having the inventory, such as the manufacturing lead time or constraints in manufacturing processes must be made apparent. The manufacturer can then reduce the limited inventory toward the zero mark by eliminating these reasons (i.e., causal factors).

For example, a supplier can continually reduce its limited inventory in pursuit of the zero inventory goal by making various improvements, such as shortening the manufacturing lead time between receiving buyer orders and delivering the ordered products, and establishing manufacturing process conditions (smaller lots, zero defects, higher production capacity, etc.) that enable production to be accurately coordinated with product lots to be delivered to buyers. Again, the ultimate goal of the ESP Production System is to maintain zero inventory, which means the supplier needs to eventually eliminate even a limited inventory.

Kanban and in-process inventory

Now let us look at why a kanban system requires a certain amount of in-process inventory, also known as inventory store, which is further described later on in *Supplemental note 2*. Kanban's in-process inventory should be thought of as a last resort that is used when all production processes are not being controlled tightly enough. It is also important to keep the kanban system in perspective as a system that resulted from many years of innovative improvements devised by an automobile manufacturer.

Consider for a moment what an automobile manufacturer's production line is like. To begin with, its production line is huge and complex. The main production flow for automobiles goes from the press line to the car body line, coating line, and assembly line. Many sublines, such as the engine assembly line, instrument panel assembly line, and seat assembly line, feed into this main line, and there are even sub-sublines that feed into the sublines. The automobile manufacturer's main production line consists of groups of linked production processes. (Its production line, as it has been described so far, does not include either the production processes of suppliers or warehouses for distribution.)

This gives us some inkling of just how difficult it must be to exercise unified control over such huge and complex automobile manufacturer production lines. Also, consider that when the final assembly schedule (specifications) is sent to the main assembly line, at that point the car bodies have not yet been fed to the main assembly line. This means that it is not possible to base production at upstream processes on the final assembly schedule. This means it is not only extremely difficult to exercise unified control over the in-process inventory among the automaker's production lines, it is logically impossible.

Kanban was devised as a means of minimizing inventory under these conditions. Since the production line cannot be controlled in a unified way, the kanban system uses functions, that might be compared to the autonomic nervous system, at individual processes, and some of these functions involve methods for controlling in-process inventory at each process. To enable production under these conditions, automobile manufacturers cannot avoid maintaining some in-process inventory between main lines, as well as at the ends of sublines and sub-sublines.

In any case, the adage that inventory is evil still holds painfully true among manufacturing managers. But it should be pointed out that *not all inventory is evil*. The evil kind of inventory is the kind that is not being controlled. Having uncontrolled inventory can lead to serious errors in judgment, such as:

- Decisions leading to a worsening of cash flow and other financial problems.
- Poor choices of areas for investment, based on misunderstanding of the company's strengths and weaknesses.
- Slowness in making urgently needed improvements of work processes or production processes during the stages from development to design, order reception, procurement, production, and delivery.

Of course, some of the above problems (a worsening of cash flow is one) can still occur even when all inventory is controlled inventory. Nevertheless, if all inventory is controlled and the relevant managers have an accurate understanding of current inventory levels, they can take effective measures against problems in their company's work processes or manufacturing processes before the problems become too serious. They can also invest wisely, based on an accurate assessment of the company's strengths and weaknesses.

Meanwhile, factory-floor managers and operators already understand that inventory is bad. They know from experience that, for example, having loads of finished products, semifinished products, parts, and materials lying around the factory causes time to be wasted on searching for things, replacing things, and moving things around. They also know that safety problems arise more easily when unnecessary stuff clutters up the factory. And they know that when things are left lying around long enough, they can become obsolete or otherwise unusable.

However, maintaining some amount of inventory to enable flexibility in the face of the unexpected, such as last-minute requests from buyers or downstream processes, or varying availability of production resources (personnel, equipment, parts, and materials) in the company's own manufacturing sections and processes, provides factory-floor managers with the means to carry out their responsibilities. That is why some amount of inventory is generally tolerated.

At some factories, the authors have seen shops where desk drawers are crammed full of old, soon-to-be-obsolete parts, just sitting there—all ready in case someone asks for one of them. In fact, we have come across just this type of scene at many different kinds of companies and industries. (The people who showed us their crammed-full desks were very candid in doing so, or perhaps they were just desperate for desk space.)

Being able to reliably control inventory means understanding one of the biggest causes of poor business performance, and it not only boosts the efficiency of factory-floor operations but also spares factory managers and operators a lot of grief.

Supplemental note 1: Why finished products are kept as limited inventory. One of the most basic issues you deal with when designing a production management system is that of where to put the inventory (i.e., "where are the stock points?"). In this book, whenever (limited) inventory is being discussed as part of the ESP Production System, it should be understood that this means inventory of finished products. Before we actually adopt and deploy the ESP Production System, we conduct a study of the company's product features, production process characteristics, handling of received orders, and so on, so that the company can determine the correct stock points for various parts of the inventory.

For example, if the company has several product models whose specifications vary only slightly and the finished products of each model are separated according to their labels or printed information, then the limited inventory of such products should be placed immediately before the process where these products will be labeled or printed.

One of the main elements in the ESP Production approach that the reader must keep in mind, is the basic rule concerning stock points where inventory is retained under the ESP Production System: the inventory is always finished product inventory. That is, if production (process, assembly, molding, or alteration) at the previous process is entirely for semifinished goods, while production at the next process is entirely for finished goods, and the next manufacturing process is one where various specifications will be built into products, then the semifinished goods from the previous process are not automatically retained as in-process inventory. Instead, only the number of products required for parts in finished products is retained as semifinished goods from the previous process. (This will be discussed further in Chapter 4.)

Naturally, after you have completed the deployment of parts for the finished goods, there is still the separate issue of establishing equalized units or ESP production patterns for the semifinished goods from the previous process and the finished goods at the next process. Equalized units and ESP production patterns are explained in more detail later in this chapter.

Again, a basic principle of ESP, is the idea that semifinished products from the previous process should not be retained as in-process inventory, but rather should be treated as finished goods. Consequently, the manufacturing process improvements you make to implement this principle are the ESP Production System's most important themes. In exceptional cases, such as when there is an extremely large amount of similarity among the previous process's semifinished products or when, despite the manufacturing process improvements made at the previous process, the previous process has an extremely short production lead time or a long changeover time that precludes the handling of small lots, the previous process is regarded as a supplier that provides parts and materials. Therefore, you would use a kanban system to withdraw semifinished goods from the next process.

In any case, ESP does not have a standard form, but rather involves making manufacturing process improvements that take into account each company's own characteristics. This is why the idea that limited inventory consists only of finished products is one of the most important principles to be conveyed when describing the ESP Production System's approach and methodology.

Synchronized and Equalized Production

Synchronized and equalized production, which are methods used to shorten lead times and reduce inventory, are the most important features for establishing the ESP Production System and realizing its benefits. Figure 3-3 shows an overview that will help to define these two key features.

Synchronized	Synchronized purchasing	Synchronized acceptance of parts and materials from vendor.
	Synchronized manufacturing processes	Synchronized manufacturing processes at own company: • Virtually no in-process inventory.
	Synchronized with buyer's delivery deadline	Synchronized with buyer's delivery deadline (synchronization of shipment and delivery deadlines).
Equalized	Equalized purchasing	Acceptance of equalized parts and materials from vendor: • Delivery units of parts and materials from vendors are coordinated with production units.
	Equalized production units	Equalized production units for each product item number under the ESP production schedule: • Set production output per item number and then establish equalized sizes. • Use equalized units in production specifications. Do not use fractions in specifications. • Units for "produce ←→ do not produce" and "stops and insertions."
	Equalized supply parts that are coordinated with pro-duction units	Units of transfer for goods (parts, materials, semifinished products) at manufacturing processes.

Figure 3-3. Overview of Synchronized Production and Equalized Production

The concept of synchronization is widely known and used, such as when it is part of a JIT Production System. Under ESP, synchronization remains similar to how it is generally understood. Synchronization is very important for any company that adopts the ESP Production System, and it must be made very specific. It goes without saying that, in order for synchronization to create an efficient cash flow, its goals must be to synchronize ordering, production, and delivery only of needed items at the needed times and in the needed amounts.

One of the methods used to achieve synchronization is the kanban system, frequently used as part of the JIT Production System. This is where the locations of inventory (of finished products, in-process inventory, parts, and materials) and the amounts (store inventory) at each location are established so as to make the inventory clearly visible. Making inventory levels visible enough to be understood at a glance helps promote the improvement of processes. In addition, the improvement effects of shortening lead time, shrinking lot sizes, and curtailing defects also become visible in the form of reduced inventory (including in-process inventory).

One of the collateral benefits of promoting process improvements based on actual factory conditions, actual equipment and materials, and actual facts is that the particulars of each process become more visible, which raises the control levels in factories, in terms of productivity, quality, cost, delivery, safety, and morale.

You can gain similar functions and effects under the ESP Production System. In particular, you can achieve three forms of synchronization: 1) synchronized purchasing, 2) synchronized manufacturing processes, and 3) synchronized delivery to buyers. Under ESP, the Planning Department performs a simulation based on an ESP production pattern to reliably estimate the required amounts of production, then deploys these required amounts using the ESP production schedule to come up with production specifications (for purchasing, production, etc.) for each process using equalized units per product item number (semifinished and finished products), which involves planning strict synchronization so as to eliminate inventory between processes.

Since this means that, under ESP, purchasing and production are carried out push-production style based on specifications issued by the Planning Department, as a rule all inventory has strings attached in the sense that each item of inventory has an assigned destination. As a result, inventory control is even tighter under this system than when using a kanban system.

The ESP Production System also outperforms the kanban system in the way it clearly emphasizes the places where manufacturing process improvements (based on actual conditions, materials, and facts) are needed and in the way it makes both the process and the results of manufacturing process improvements clearly visible. The upshot is that the ESP excels in raising factory-floor control levels in terms of productivity, quality, cost, delivery, safety, and morale.

Supplemental note 2: Why ESP has tighter inventory control. Since kanban is a revolving type of system in which you replenish used kanban, some unassigned inventory (inventory with no assigned use or destination) is retained to guarantee a supply during the period between the withdrawal and replenishment of kanban. In other words, the kanban system is a pull-method system in which items are withdrawn by downstream processes. A store inventory is retained to guarantee a supply of goods between the time when the next process issues a kanban order and the time when the previous process (or the outside vendor) provides replenishment of the withdrawn items. The items in this store inventory are unassigned items.

When using kanban, it is possible to prescribe levels of store inventory so as to control it to some extent, but it is not possible either to make the inventory items assigned items or to maintain zero inventory. In this sense, the ESP Production System provides tighter inventory control.

Synchronization of purchasing

Synchronization of purchasing means synchronizing the acceptance of parts and materials from vendors with the production schedule. Under ESP, you achieve synchronization of purchasing mainly by deploying and ordering parts and materials based on the confirmed ESP production schedule, which is created at the stage where you issue production specifications to the factories. However, when ordering parts and materials based on a confirmed ESP production schedule, it is not always possible to meet the required lead time for purchasing. In such cases, you use ESP production patterns as a basis for deploying and ordering parts and materials.

There are generally four methods for purchasing parts and materials (see Figure 3-4). The ESP Production System basically uses the periodic reordering of variable amounts method. However, even though we say variable amounts, under ESP all product item numbers have equalized production units and are organized into patterns, so that you can determine equalized units for purchasing for all parts and materials to be purchased. All orders made under the ESP Production System are made as multiples of these equalized units for purchasing. This enables the supplier to provide more precise in-house information when contacting vendors and helps the supplier prevent missing parts and materials, which also means that the supplier can be more effective when negotiating for price discounts with vendors, which becomes a means of cost reduction.

Amount ordered \ Ordering interval	Periodic	Indefinite
Fixed amount	Fixed amount in set period. Rarely used in manufacturing	"Double bin" method, etc. Method used when necessary by the ESP Production System and by the kanban system
Variable amount	Method basically used by ESP Production	Method basically used by kanban system

Figure 3-4. Four Purchasing Methods

Depending on the characteristics of the parts and materials being ordered and/or the vendors being used, you can make purchasing more effective by using a kanban system. In terms of characteristics, parts and materials can be broadly divided between parts and materials corresponding to product item numbers (no commonality with other products), and extraneous items (parts and materials) that remain after deployment of required amounts according to the confirmed ESP production schedule. Examples of the latter might include coating materials, product labels, and casting materials. Even under the ESP Production System, you can use kanban for purchasing these types of items (parts and materials).

In addition, large-scale buying (such as when using a periodic fixed amount method) may be the best option for purchasing inexpensive parts and materials that are used in large amounts. For example, such parts and materials might include bulk items such as nuts and bolts. In any case, it is only natural that you should apply different purchasing methods for different items and vendors once you have studied and categorized the characteristics of the various parts, materials, and vendors. The key point to bear in mind is that under ESP, major parts are as a rule ordered via the periodic indefinite amount method, based on the confirmed ESP production schedule.

Synchronization of manufacturing processes

Synchronization of manufacturing processes means strictly synchronized production based on the ESP production schedule, which the Planning Department has confirmed and has communicated to the relevant production processes (major processes where production specifications are sent). Under ESP, production is based on the ESP production schedule, in which both the production sequence and the production output amounts are specified. In other words, since specifications are issued using a push method from the Planning Department to the Production Division, all inventory at manufacturing processes is assigned inventory (inventory with strings attached). And since the production specifications clearly indicate not just the production amount but also the production sequence, full synchronization becomes possible for manufacturing processes.

Synchronization with buyer's delivery deadline

Synchronization with the buyer's delivery deadline is significant as a way to prevent missing items in deliveries. As was mentioned in the first section of Chapter 3, when a supplier implements synchronization with the buyer's delivery deadline, it is better able to ensure its Zero Missing Products Guarantee, which is of utmost importance in maintaining trust among buyers. When a company implements ESP correctly, it is able to respond to orders from buyers by delivering the needed parts in the needed amounts before the delivery deadline.

Equalized Amounts

Equalized amounts is an original concept in the ESP Production System. It means using identical quantities when ordering or receiving purchased goods, manufacturing products (semifinished or finished) based on production units (amounts) per product item number, or supplying parts and materials in coordination with the production units (amounts) per product item number. Equalizing makes it possible for a supplier to be more specific about its Six Guarantees by greatly reducing the volume management workload of the Purchasing Division or production process supervisors. In fact, it is no exaggeration to say that any company that firmly practices the concept of equalizing has managed to establish something like an ESP Production System.

Under ESP, when production units for specific product item numbers (of semifinished or finished products) have been equalized, you can draft an ESP production schedule by simply deciding whether or not to produce certain, predefined amounts. This means that if the production unit for item No. A is 100 (i.e., item No. A is always produced in 100-piece lots), the decision is always a matter of deciding between manufacturing 100 of that item or none at all.

The method used to make this type of decision is to first check the information on received orders and inventory in the ESP production patterns, and then to revise (arrange) the sequence of target items as needed. Using the ESP production patterns to revise (arrange) items in this way helps to further solidify the confirmed ESP production schedule.

Another way to describe this revision method is to compare it with the binary system (of 1s and 0s). When you incorporate a binary approach into the concept of equalizing with regard to planning and proposing production schedules, you can support the Six Guarantees more rationally and computer-based simulations become possible. In fact, equalizing is essential when seeking to replace the old inner-circle approach to production management with a production management approach with a more rational and transparent structure.

Under ESP, equalizing happens in three ways corresponding to 1) purchasing, 2) production units, and 3) supply of parts coordinated with production units. We'll now discuss each of these.

Equalizing purchasing

To equalize purchasing, purchasing amounts, called equalized purchasing units are set for each item number (or item name) of parts and materials corresponding to the parts and materials that are delivered by vendors. Purchase orders are then always issued using multiples (x1, x2, x3, etc.) of these equalized purchasing units.

The equalized purchasing units for parts and materials (i.e., units set for each item number of parts and materials) are set according to the results of required amounts of parts and materials, which have been determined based on the equalized units (amounts) of finished products at the company and the required amounts of parts and materials deployed from ESP production patterns. The following benefits can be expected to come from equalizing purchasing.

- Simplification of ordering tasks.
- Simplification of quantity checks when receiving ordered goods.
- Simplification of inventory management for parts and materials.

Equalizing production units

Equalizing production units means setting production units (amounts) for each product item number (of semifinished and finished products) and using a multiple (x1 or greater) of these production units to set amounts for production. All production specifications are made using multiples of production units. If, during the ESP production schedule planning process, it is determined that certain (semifinished or finished) products do not need to be manufactured, then the production specifications for those products will show "0" instead of a multiple value. In other words, the production specifications will either show "0" or a multiple value (x1 or greater) for each equalized unit.

Likewise, under ESP, you use equalized units when specifying the stops and insertions that are used to arrange ESP production patterns. The equalized unit for each product item number is set after surveying and analyzing periodic (such as monthly) totals of orders received, as well as delivery cycles and constraints on manufacturing processes. (For details, see Chapter 4.)

One basic caution regarding equalized units for specific product item numbers is that lot sizes should be small. If equalized units are for large lots, it will be impossible to fully meet the Six Guarantees no matter how well you have established the ESP Production System in other respects. The following benefits can be expected to come from equalizing production units.

- In principle, there is no longer any need to check quantities at the company's in-house manufacturing processes (intermediate processes). Consequently, it means less factory management work, such as collecting and analyzing data.

- The quantities to be found within each manufacturing process can be understood simply by consulting the number to be fed to the initial process and the number to be output from the final process as specified by the Planning Department.
- In principle, the Planning Department has a clear understanding of the amounts specified for production (amounts to be supplied to the production lines), so there is no need for each process (in the factory) to keep tabs on the numbers of supplied items and nondefective items.
- If process (factory) managers ever need to check quantities, they can simply check the number of defectives at each process to get a precise figure for quantities being managed at each process (factory).
- Likewise, the transit information from each process can be used to manage progress in relation to the production schedule. This makes process management an easier task.

Equalized supply of parts coordinated with production units

Equalized supply of parts coordinated with production units means that all parts and materials required for the production processes of semifinished and finished goods manufactured by the company can be supplied when they are needed and in exactly the needed amounts. In other words, the parts and materials needed for production of the supplier's products have been coordinated with the production units (equalized units) so that only the required items in the required amount will be delivered at the required times.

Or, from a different perspective, the equalized supply of parts coordinated with production units means that transfer units exist for the supplier's goods (including parts, materials, and semifinished products). An equalized supply of parts coordinated with production units is based on the equalized units for each product item number described above and are used to deploy and establish the required amounts of parts and materials.

The optimum situation is when the equalized units of parts to be supplied are equivalent with the equalized units for purchasing. When these two units are not equivalent, the equalized units for purchasing must be set as an integral multiple of the equalized units of parts to be supplied. The following benefits can be expected to come from having an equalized supply of parts coordinated with production units.

- Simplification of tasks performed by process managers and parts/materials transport managers.
- Elimination of the need for setup operations (such as counting or arranging parts in warehouses) for supplying parts and materials to assembly lines.
- Elimination of the need to identify and manage fractional sets of parts and materials.
- Elimination of the need to rearrange parts and materials or perform checks to prevent supply errors when changes have been made in item numbers used in production of the company's own products.
- Simplification of checking procedures by process operators.
- Elimination of the need to check for causes of missing parts and materials, such as when too few or the wrong kind of parts or materials have been supplied.
- Elimination of the need to cross-check the production output plus the amount of surplus parts and materials when changes have been made in item numbers used in production of the company's own products.
- Elimination of the need for operators to perform checking operations such as checking for missing or surplus assembly components, incorrect parts and materials, or incorrectly assembled or processed parts and materials at their processes.

STEPS FOR REALIZING THE SIX GUARANTEES USING ESP PRODUCTION PATTERNS

Now that we have provided an overview of the Six Guarantees and how these six steps are used in computer simulations to build an ESP Production System, as well as outlining the basic steps for implementing ESP, we can now discuss the role of ESP production patterns, as determined by the Six Guarantees, in ensuring the success of the ESP Production System.

Role of ESP Production Patterns

Through the use of simulations and computers, the ESP production pattern serves as the basis for methods for simplifying the development of an ESP Production System. As we mentioned, you can define the ESP production pattern as an early, preliminary production plan that helps the company clarify its production planning standards, as well as anticipate demand from buyers. Another way to think of the ESP production pattern is as a type of substitution formula—the formula being an expression of the early, preliminary production plan. Using a substitution formula makes it easier for you to run your simulations. It also enables you to run successive simulations according to the timing by which various buyers send in orders, including preliminary in-house order information and confirmed order information. You adjust (rearrange) the substitution formula as this order information comes in.

Thus, the supplier makes these adjustments to their substitution formula, i.e., the ESP production pattern, and when the time comes to issue a production schedule for its production processes, it is able to issue an ESP production schedule (confirmed schedule) whose production deadlines are based on its own planning decisions rather than being directly based on the influx of orders. Figures 3-5 and 3-6 describe the role of ESP production patterns.

How does the ESP Production System utilize ESP production patterns?

An ESP production pattern is a type of substitution formula that we use to analyze scenarios based on proposed manufacturing of goods.

Advantages of using a substitution formula include:
- It facilitates simulations made in relation to the Six Guarantees.
- It helps the supplier switch from passive/conservative production planning to dynamic/progressive production planning (see Figure 3-6).

Caution points concerning an ESP production pattern include:
- It is nothing more than a substitution formula.
- It must be adjusted (rearranged) as a substitution formula.
- After the adjustments are completed, it must be issued as an ESP production schedule (i.e., a confirmed production schedule).

The ESP production schedule, which utilizes ESP production patterns, is the basic principle of production scheduling.
- It helps planners create more accurate and specific production schedules.
- It helps prevent the need for adjustments to the production schedule after it has been issued.

Figure 3-5. Role of ESP Production Patterns

ESP Production Schedule	Conventional Production Schedule
ESP production patterns have been proposed in advance. Adjustments (rearrangements) are made each time new order information (including in-house order information and confirmed order information) is received from buyers. The supplier can issue confirmed order information (the ESP production schedule) up to the last minute possible for the target production processes and can fine-tune its scheduled production periods. Once the ESP production schedule has been issued, it is not changed.	Production schedule is proposed after order information (mainly confirmed order information) has been received from buyers. Each time confirmed order information is received from buyers, new production scheduling specifications or a new confirmed production schedule (ESP production schedule) is sent to the company's production processes. Once the confirmed production schedule has been issued, if there are any subsequent changes in buyer orders, a "revised production schedule" must be issued for each set of changes.
Dynamic (progressive) production planning	**Passive (conservative) production planning**

Figure 3-6. Comparison of ESP Production Schedule and Conventional Production Schedule

Overview of ESP Production Patterns

The model in Figure 3-7 provides a model of ESP production patterns. As you can see, you establish a separate ESP production pattern for each production process or production line where production specifications are sent.

You determine each ESP production pattern based on four criteria, which in turn are based on the planning standards. These four criteria make simulations both easier to perform and more reliable, and help to establish a rational production planning that lends itself to computerization while weaning the company away from decision-making by an inner circle of managers.

"Firming-up" of products in production processes (lines)

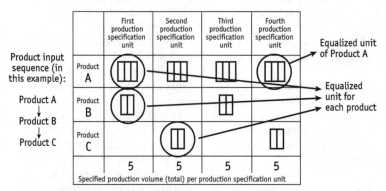

Figure 3-7. Model of ESP Production Patterns

1. *Criteria 1: Establishing production processes (lines) for specific product item numbers to stabilize production items and production processes.* Ordinarily, the conditions for processing and manufacturing differ for each product item number and the available production processes (lines) are therefore limited. When using ESP production patterns, a supplier establishes a separate production process (or line) for each product item number and this is firmed up during the period when the ESP production patterns are subject to adjustments. Often, this firming up is done not so much for individual product item numbers as it is for groups of products. Occasionally, as an exception, prioritization is set among a number of production processes (or lines) that are able to handle certain product item numbers. Of course, the more generalized production processes that are able to handle almost any type of product are not subject to firming up.

2. *Criteria 2: Establishing input sequences for specific product item numbers.* The company establishes an input sequence (a product manufacturing sequence) that is based on the Production Efficiency Maximization Guarantee, (guarantee four discussed earlier).

3. *Criteria 3: Establishing equalized production units.* Equalized units are established for specific product item numbers as units

used in simulations and production specifications (discussed earlier, as well as in Chapter 4).

4. *Criteria 4: Establishing specified production volume for each production unit as 100 percent of the production capacity.* All ESP production patterns must be designed to fulfill the Production Efficiency Maximum Guarantee. Therefore, using the criteria of input sequences and equalized units, the total amount of specified production (i.e., the production load) is established so that it equals 100 percent of the production capacity. This total production load is comprised of various production specification units (such as production specifications made for time periods of 30 minutes, 2 hours, per shift, per day, every two days, or per week).

Key Perspectives When Establishing ESP Production Patterns

It is important to take an approach from one of the following two perspectives when performing studies and analyses prior to establishing ESP production patterns for locations that receive production specifications, such as specific production processes or lines.

The first perspective is to clarify your company's production planning standards. When you make improvements in your manufacturing processes, you must conduct studies and analyses to determine whether or not your improvements require revision or further clarification of the planning standards used for production planning. (See *Supplemental note 3*.) The following three points summarize the planning standards that are inevitably needed in order to establish ESP production patterns.

1. Firm up the production locations (lines, processes, equipment, etc.) for each product item number, including semifinished and finished products.
2. Establish production load standards (such as production time per item) for each product item number, including semifinished and finished products) and, if necessary, for each production location.
3. Establish an input sequence at each production location for each product item number, including semifinished and finished products.

Supplemental note 3: Topics of study and analysis concerning planning standards. First you look at the topics of study and analysis concerning manufacturing processes.

- Product item number lists, process analysis, and analysis of work operations and methods.

- Loss analysis (equipment losses among equipment, labor losses among personnel).
- Distinctions between in-house and outsourced operations.
- Processing categories and sequence of machining processes.
- Sequence of assembly processes and distribution of instructions, etc.

Then you look at the topics of study and analysis concerning planning standards

- Configuration of parts
- Defect rate (yield rate)
- Changeover labor hours
- Daily scheduling standards
- Load standards
- Capacity standards
- Inventory standards . . . etc.

The second perspective is to take into account the situation regarding orders received from buyers. One way is to estimate future orders based on past orders, such as during the previous half-year, year, or two years. Another way is based on estimating supply and demand predictions. No matter which method you use as the main method, you must also consider the buyer's policies on sales, production, and ordering.

However, when a supplier uses supply and demand predictions (i.e., statistical methods), whether it is an analysis based on statistical methods or a study of the results of such an analysis, the supply and demand predictions will not be very reliable unless you eliminate factors that cause artificial fluctuation in supply or demand. (See *Supplemental note 4.*) Consequently, in many cases, orders from buyers are predicted based on a combination of studying past orders and considering the current trends and ordering policies of final companies. Supplemental note 5 describes the topics of analysis used when studying past orders and estimating current conditions among buyers.

You need the following two planning standards when establishing ESP production patterns based on estimated ordering by buyers.

1. Establish equalized units for specific product item numbers, that is, semifinished and finished products.
2. Establish limited inventory standards for specific product item numbers, that is, semifinished and finished products.

Supplemental note 4: Main factors that cause artificial fluctuation in data. When you analyze your buyers' data you'll notice two factors that causes it to artificially fluctuate.

1. The precision of the order data (product models ordered, quantities of each model, etc.) can vary greatly depending on the type of production management system used at each buyer.
   - Production management systems are generally divided into two types: 1) order-based production management systems and, 2) planning/estimate-based production management systems.
   - The planning/estimate-based production management systems are further divided into two types: 1) production management systems based on supply and demand predictions and sales plans, and 2) production management systems based on maintaining inventory levels.
2. When a buyer purchases goods from several suppliers, the types of goods it orders from specific suppliers may change whenever the buyer changes its purchasing policies.

When you analyze your own company's production plans and production results you'll notice two factors that cause the data to artificially fluctuate.

1. If your company's Production Management Division still relies on an inner circle of experts to carry out its planning tasks, the order data it receives from buyers may be interpreted according to the experience-based rules of this inner circle, or it may become otherwise clouded by their particular opinions and judgments.
2. Sometimes the data includes requirements and requests from the Production Division.

Supplemental note 5: Topics of analysis used when studying past orders and estimating current conditions among buyers.

- Analysis (such as ABC analysis) of ordered amounts per product item number.
- Analysis of distinctions between in-house order information and confirmed order information.
- Analysis of variation modes (constant variation, trends, seasonal variation, regular variation, irregular variation, exceptional variation, etc.).
- Analysis of monthly (or semimonthly) averages.

- Analysis of daily averages.
- Analysis of delivery units, for example, the number of items covered by each delivered kanban.
- Analysis of likely future trends, etc.

Process of Proposing and Specifying an ESP System Based on ESP Production Patterns

Figure 3-8 illustrates the process whereby the ESP production patterns are used to propose and specify an ESP Production System that fulfills the Six Guarantees. It does this while being both fine-tuned for scheduling efficiency and not subject to change once

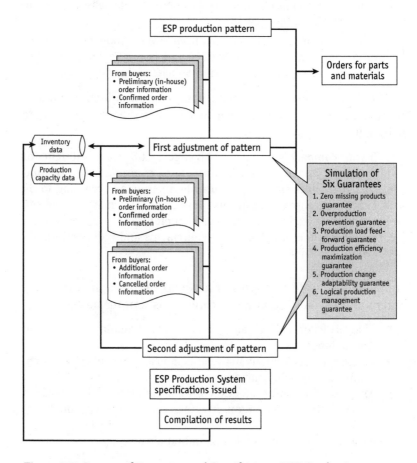

Figure 3-8. Process of Proposing and Specifying an ESP Production System Based on ESP Production Patterns

issued. In Figure 3-8, the ESP production patterns are adjusted (rearranged) twice. This illustration simply provides an image of where such adjustments might be included. In actuality, they occur as necessary according to each company and/or production process.

Let us now examine a simplified example of the steps and timing by which a supplier issues its ESP Production System specifications. We will assume that the manufacturing company in this example, X Industries, is working from the received order information that is listed in Figure 3-9.

Order information from Company A:
- Confirmed order information is received on a daily basis.
- Preliminary order information for the next two months is received each Friday.

Order information from Company B:
- Confirmed order information for the next 10 days is received every 10 days.
- Preliminary order information for the next three months is received on the 10th of each month.

Figure 3-9. Example of Timing of Receiving Order Information at X Industries

From Figure 3-9, we can see that at X Industries, the ESP production patterns are affected by confirmed orders received daily and preliminary orders received each Friday from Company A. In addition, they are also affected by confirmed orders received every 10 days and preliminary orders received on the 10th of each month from Company B. When X Industries receives this order information, it works the new information into the substitution formula (intrinsic to ESP production patterns), performs simulations to ensure the Six Guarantees, and then adjusts (rearranges) its ESP production patterns accordingly.

Supposing that X Industries issues specifications to its production processes each afternoon for the next day's production schedule, it would therefore be issuing a (confirmed) ESP production schedule that is a daily schedule issued the previous afternoon (see Figure 3-10). This figure helps illustrate the production planning principles of making last-minute fine-tuning adjustments and preventing changes in confirmed schedules once they are issued.

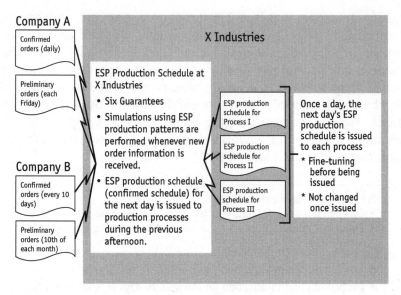

Figure 3-10. Example of Operations at X Industries

Image of How a Supplier Adjusts Its ESP Production Patterns

Figure 3-11 shows how a supplier adjusts or rearranges ESP production patterns. First, the supplier must incorporate the order information (preliminary and confirmed) that it receives from buyers into its ESP production patterns, such as by calculating production, shipment, and inventory amounts for each product. (Figure 3-11 shows a basic production pattern.) The *Inventory trends* column on the right side of Figure 3-11 shows graphed results from inventory calculations for the basic production pattern. In this example, these results include the following information.

- If Product A is manufactured according to the basic production pattern, there will be no missing products and appropriate inventory levels will be maintained.
- If Product B is manufactured according to the basic production pattern, overproduction will occur and the inventory, at unit 4 of the range used in the latest schedule simulation, will exceed the limited inventory level.
- If Product C is manufactured according to the basic production pattern, missing products will occur at unit 2. Furthermore, at unit 4 of the range used in the latest schedule simulation, the limited inventory standard will be violated, making it impossible to ensure adequate inventory levels.

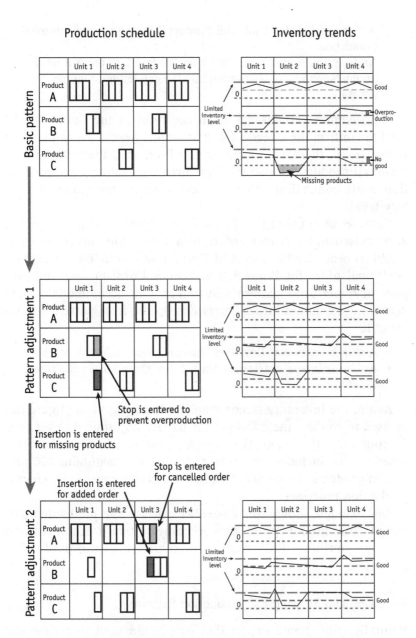

Figure 3-11. Image of Adjustments Made in ESP Production Patterns

Using these results, the company performs pattern adjustment 1. These results are illustrated in the middle part of Figure 3-11, indicated as *Pattern adjustment 1*.

- A stop is entered at Unit 1 for Product B in order to prevent over-production.
- An insertion is entered at Unit 1 for Product C to prevent missing products. (Product C is inserted into the gap left by stopping Product B.)

The inventory trends graph on the right in the figure can be used to confirm the effects of pattern adjustment 1. As can be seen in the graph, there are no longer any Product C shortages at Unit 2 and all products can be manufactured without any overproduction while controlling inventory levels within the limited inventory level.

Next, let us suppose that just before starting production of Unit 2, a revised order is received from a buyer. The buyer has cancelled its order for Product A at Unit 3 and would like to make an additional order for Product B at Unit 3. Consequently, the supplier next performs pattern adjustment 2 to accommodate these last-minute changes into the previous results from pattern adjustment 1.

- A stop is entered at Unit 3 for the cancelled Product A order.
- An insertion is entered at Unit 3 for the additional Product C order.

Again, the inventory trends graph on the right in the figure can be used to confirm the effects of the pattern adjustment. After confirming the adjustment, the company drafts an ESP production schedule that includes zero overproduction and maintains 100 percent of production capacity, then issues this schedule to the relevant production processes.

This example shows how a supplier can use ESP production patterns to fine-tune its production schedule so that it can devise and issue an optimum ESP production schedule to the production processes.

Reviewing and Revising ESP Production Patterns

Naturally, you should expect that you would need to review and revise the ESP production patterns, since the content and amount of orders received from buyers can change. This will occur when the number of buyers changes (through gain or loss of buyers) or when the types of products ordered change, due to higher or lower orders for certain product item numbers, design changes, etc. A supplier must also update its ESP production patterns after it installs new or improves old manufacturing processes, or when it changes its production capacity (which means changing the planning standards).

As we mentioned in our discussion of Guarantee Two (Production Load Feed-Forward Guarantee) at the beginning of this chapter, you must set and follow rules with regard to how and when you should review ESP production patterns. Generally, each ESP Production System includes a set of rules specifying the functions for gathering and analyzing order information and the timing (monthly, semiannually, etc.) of ESP production pattern reviews based on such information. These functions deal not only with order information but also ESP scheduling information, production result data, and other types of information. Such information is analyzed and reviewed to support ongoing improvement of ESP production patterns.

In addition, ad-hoc reviews of ESP production patterns are also necessary whenever a warning occurs concerning standards or rules that are no longer being followed. In cases where you cannot continually use the same production pattern, the supplier can choose to establish ESP production patterns that are used for only a specified period of time.

Steps to Expand Production Capacity and Create Flexible ESP Production Patterns

As was mentioned in our discussion of the Four Principles of the ESP Production System in Chapter 2, ESP cannot succeed unless it includes improvement of manufacturing processes. Therefore, operating the ESP Production System using ESP production patterns is premised upon the following two points.

1. The production capacity gained by using ESP production patterns is at least enough to cover the amount of orders received (i.e., the workload).
2. Product categories or item numbers are firmed up or grouped for specific production processes (or lines) in order to simplify and clarify the use of ESP production patterns and the management of production operations.

Consequently, a supplier should view as essential any technique that supports the expansion of production capacity and the establishment of flexible ESP production patterns. It is only natural then that as you continually raise the level of manufacturing process improvements, you are also able to create ever more effective ESP production patterns. In fact, these improvements become a prerequisite for successfully managing ESP. Here are three key points to consider in relation to improving manufacturing processes.

1. Shorten lead times.
2. Reduce intermediate (in-process) inventory to zero or almost zero.
3. Overcome conditions that put constraints on manufacturing.

The four points listed in Figure 3-12 are techniques (approaches) that embody the above three points related to improving manufacturing processes.

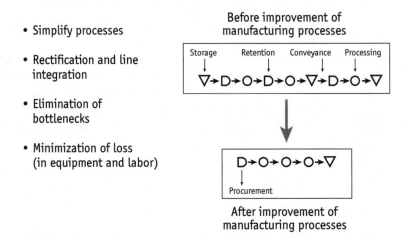

- Simplify processes

- Rectification and line integration

- Elimination of bottlenecks

- Minimization of loss (in equipment and labor)

Figure 3-12. Methods for Achieving Improvement of Manufacturing Processes in an ESP Production System

When implementing these manufacturing process improvements, be sure to apply the ECRS principles of improvement (Eliminate, Combine, Rearrange, Simplify), described in Chapter 2, to maximize the improvements.

For its part, the Production Management Division should provide leadership to the Manufacturing Division by presenting specific sites, topics, and orientations in need of improvement, thereby helping to direct the making of manufacturing improvements in the interest of making the company's ESP Production System the best it can be.

For further elaboration of the manufacturing process improvement goals and approaches, see "Four Principles Behind the Construction and Operation of ESP" in Chapter 2, and for a description of the four techniques used to carry out manufacturing process improvements, see "Key Points for ESP Production System Innovations" in Chapter 4.

BEFORE AND AFTER BENEFITS OF THE ESP PRODUCTION SYSTEM

Let us sum up what we have covered so far in this chapter by comparing the situation at a company before and after introduction of the ESP Production System. Figure 3-13 illustrates this comparison.

Let us go step by step through the comparison that is illustrated in the figure.

1. ESP is a synchronized push production system in which previously equalized units for each product item number are specified to the first processes.
   - Synchronize production to minimize in-process inventory, including inventory found between equipment units or between manufacturing shops.
   - Seek the maximum possible reduction in lead time.
   - Thoroughly apply the ECRS principles as they relate to process design.
   - Eliminate processes, combine processes, rearrange processes, and simplify processes.
   - Shorten cycle times.
   - Eliminate bottleneck processes and improve equipment units prone to bottlenecks.
   - Synchronize various processes.
   - Develop synchronization among equipment units (through in-process linkage methods such as one-piece flow).
   - Develop synchronization among manufacturing shops (through linkage between processes).
   - Synchronize acceptance of materials from in-house and outsourced suppliers.
   - Thoroughly eliminate loss to help maximize the time-based operation rate and performance-based operating rate.
2. Set a limited inventory level for each product item number so that your company's production schedule is no longer directly linked to client delivery requirements.

Before introduction of ESP Production System

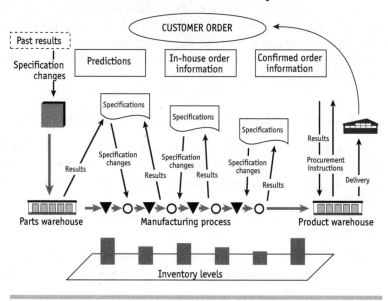

After introduction of ESP Production System

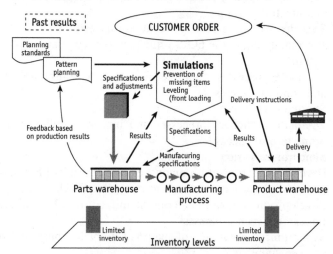

Figure 3-13. A Company Before and After Introduction of ESP

- Various products are assigned limited inventory levels.
- Limited inventory exists among products and semifinished products, but there is no in-process inventory. The purposes served by having limited inventory are:
  - To absorb variation in orders from buyers.

- To help establish an ultraefficient production schedule for the company.
- Limited inventory means that there is no inventory between processes (such as in intermediary stores), so the result is a radical reduction in overall inventory.
- Not all products (item numbers) are assigned limited inventory levels. Rather, limited inventory levels are assigned only to target products in the ESP Highway (described in Chapter 4).

3. Maintaining limited inventory is a way to ensure against missing items (while meeting all delivery deadlines) in products ordered by buyers.
4. Using ESP production patterns makes it easier to simulate the production schedule in preparation for production leveling.
5. Having an ESP information system facilitates the collection, analysis, and management of information such as production indices, yield figures, results, and numerical data.

Benefits of the ESP Production System in Addition to the Six Guarantees

The introduction and deployment of the ESP Production System can bring great success to the whole company's many functions and operations. Such results are the product of not only the ESP Production System itself, but also its underlying methods and approaches, as is summarized in Figure 3-14.

Since the results to be gained through the Six Guarantees have already been described, let us change our perspective and conclude this chapter by briefly considering six results benefiting the supplier that are not related to the Six Guarantees.

1. Results benefiting ordering and delivery operations (and results related to management of supplier companies):
   - Instead of being at the mercy of orders received from buyers (including buyers that themselves must keep pace with their own kanban systems), suppliers have enough leeway to devise the most efficient ways to conduct their production activities.
   - Strengthening the planning and systemization functions for wide-variety, small-lot production enables a supplier to respond more quickly and flexibly to production changes.
   - Suppliers that are having problems making deliveries on time can reduce late deliveries to zero by implementing proactive management. Proactive management also enables them to respond more quickly and flexibly to last-minute orders.

	Methods	Approaches	Results
Synchronization	Synchronization or procurement	Zero missing items guarantee	Zero late deliveries
	Synchronization of manufacturing processes	Overproduction prevention guarantee	Shorter lead times
	Synchronization of deliveries to buyers	Production load feed-forward guarantee	Reduced inventory and more precise inventory management
Equalization	Equalization of procurement	Production efficiency maximization guarantee	Maximization of production efficiency
	Equalization of production units	Production change adaptability guarantee	Simplification of volume control
	Equalization of supplied goods coordinated with production units	Logical production management guarantee	Simplification of specific cost control analyses
			Reduction of management costs for indirect functions
			Reduction of management costs for indirect labor

Figure 3-14. Summary of ESP Production System Methods, Approaches, and Results

- When a supplier has a production schedule that makes the most of its production capacity, it can achieve the highest possible production efficiency.
- Thorough implementation of synchronization and equalization makes effective quality control and cost control much easier to accomplish.

2. Results benefiting production management tasks (reduction of indirect labor):
   - By expanding the volume control units from parts and materials units to final product control units, production management tasks become both more accurate and less complicated.
   - Equalization facilitates (simplifies and improves the reliability of) production planning.

3. Results gained through greater computerization of production management tasks (reduction of indirect labor):
   • Enables cumbersome production planning tasks to be eliminated.
   • Use of the ESP information system automatically generates leveled production schedules.
   • Enables quick and flexible responses to changes in received orders.
   • Rescheduling also uses the ESP information system.
   • Enables high-speed processing to be done on PCs, not just the network's central computers.

4. Results benefiting process management tasks (reduction of indirect and semidirect personnel). The production sequence and production amount of each product on each day is pre-determined.
   • Enables the elimination or reduction of production follow-up operations. In ESP, the production sequence and production amounts are very clear, and once production order (production schedule) is issued, it does not change. Therefore, the follow-up operations at the production floor become much easier.
   • For the same reason, quality control operations at the production floor are also greatly simplified. This means reduction or elimination of "Poka mistakes," and indirect and semidirect personnel are able to spend more time in quality improvement operations.
   • Enables quality control (such as acceptance inspection, etc.) of parts and materials periodically and as a result, reducing or eliminating unnecessary communication/confirming operation and unnecessary inspection time.
   • Preparation works for parts and materials (such as acceptance and setting of parts and material handling within a factory) are totally eliminated or greatly reduced. Utilizing ESP makes this possible because the suppliers of the parts and materials do the preparation work. (Of course, it is a prerequisite that the parent company sets the rules, standardizes the shop-floor layout, and material handling method, etc.)
   • Because of the benefit in volume control stated below, the miscellaneous operations of indirect and semiindirect personnel are greatly reduced and simplified.
   • Enables concentration on the fundamentals of production control (man, machine, and material), such as employee training, attendance/arrangement of employees, machine/facility maintenance and improvement, grasping and improving the quality conditions of parts/materials, improvement and standardization of production method.

5. Results benefiting volume control (management of coefficients, quantities, etc.). The reduction of indirect labor and semiindirect labor:
   - Enables volume control of parts and materials at intermediary processes (in-house and outsourced) to be either totally eliminated or radically reduced, and enables simplification of tasks such as stabilizing the flow of production specifications and revisions sent to processes.
   - Implements more comprehensive volume control by going from parts and materials units to product units.
   - Since production is equalized product by product, the factory-floor work of reading and confirming production specifications becomes more accurate and simpler.
   - Equalization also makes factory-floor quality control and cost control more accurate and simpler.

6. Results benefiting inventory management:
   - Enables suppliers who have not been able to escape from having to keep too much inventory on hand (in order to be prepared for estimated production levels) to shift more toward wide-variety, small-lot production while reducing inventory level through ESP's limited inventory management approach.
   - Equalized production of the subassembly products and parts used in assembly processes means that you can eliminate shortages and excesses of intermediary parts, reducing parts inventory.

4

ESP Building Steps and Techniques

As was explained in previous chapters, the ESP Production System enables suppliers to meet the requirements of their buyers' Just-in-Time (JIT) ordering systems through wide-variety, small-lot production, while also making their own production activities as efficient as possible. In this chapter we present an overall image of ESP's architecture. To establish an ESP Production System, a company has to build up three key elements (see Figure 4-1).

1. *Synchronized and equalized scheduling:* To implement a flexible scheduling process that prevents both missing items (underproduction) and overproduction

2. *Manufacturing process improvement:* To promote manufacturing system improvements and innovations that make the production system more efficient.

3. *ESP information system:* A tool used to promptly and accurately plan schedules and to reliably administer the ESP Production System.

One characteristic of the steps for building the ESP Production System is that improvement of manufacturing processes is done concurrently with the building of the production management system. Synchronized and equalized scheduling mainly involves making improvements in the process of drafting production schedules. As was explained in the description of the Four Principles in Chapter 2,

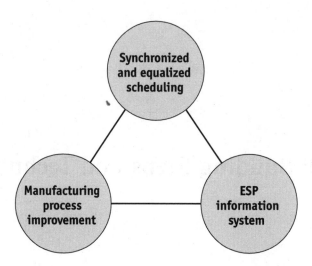

Figure 4-1. Three Elements of an ESP System

improving the manufacturing processes is an essential part of building an ESP Production System. That is why improvements in the Production Division that center on shortening lead time must be made concurrently with efforts to establish synchronized and equalized scheduling.

Development of the ESP information system does not begin until you have built synchronized and equalized scheduling. This is because the information system can be developed in a much shorter time once the system for synchronized and equalized scheduling has been established. Figure 4-2 shows an overview of the three phases for building an ESP Production System. As is shown in the figure, you can usually build ESP in about one year, but it may take longer when the target products or manufacturing processes are especially complex. Also, it is often difficult to make all necessary manufacturing process improvements and to complete the information system within one year. Consequently, during the second and subsequent years you need various follow-up activities to iron out the rough spots in the ESP Production System.

There is no need to rebuild all of the company's production management functions in order to build ESP. Figure 4-3 on page 114 shows an example of this relationship between old and new systems, in which the sections enclosed in broken lines indicate areas from the old system, such as sales planning and materials requirement deployment, that you can still use to build ESP.

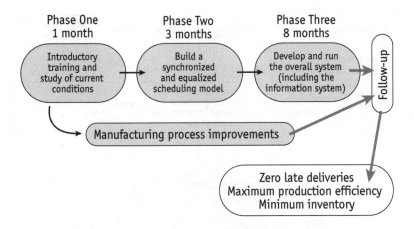

Figure 4-2. Overview of the Three Phases for Building an ESP Production System

Figure 4-4 on page 115 provides a more detailed overview of the steps for building an ESP Production System, including the main steps within each of the three phases. We will now explore those phases.

PHASE ONE: INTRODUCTORY EDUCATION AND STUDY OF CURRENT CONDITIONS

The first item of business is ESP education. Executives and other employees in the company that are responsible for building the ESP Production System teach the classes. The main topics covered are the ESP approach and the steps for building an ESP Production System. The changes that you make in the production management system will be wide-ranging and will extend to several related divisions in the company. By the same token, the people involved in an ESP Production System include not just production management staff but also people in purchasing, manufacturing, and distribution. That is why it is very important that all of the people in these relevant divisions gain an understanding and are in agreement about the production system they are going to build.

At this initial stage, the top managers need to discuss and determine their goals in building this system. If current conditions are not yet well enough understood to enable or determine these goals, they should be determined once you have completed the basic-analysis step. The steps in Phase One are mainly devoted to understanding current conditions. The objectives in doing this are:

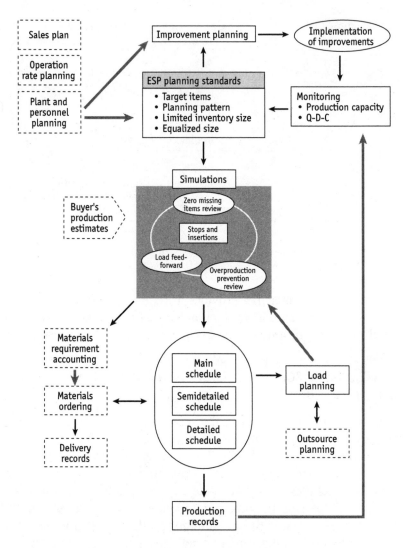

Figure 4-3. Target Areas for Building an ESP Production System

- To gain a quantitative understanding of conditions prior to improvement.
- To confirm product characteristics and manufacturing characteristics and clarify the approach to be taken when building equalized sizes and ESP production patterns.
- To determine the model to be built in Phase Two.

Gaining a quantitative understanding of conditions prior to improvement enables you to have preimprovement data (such as

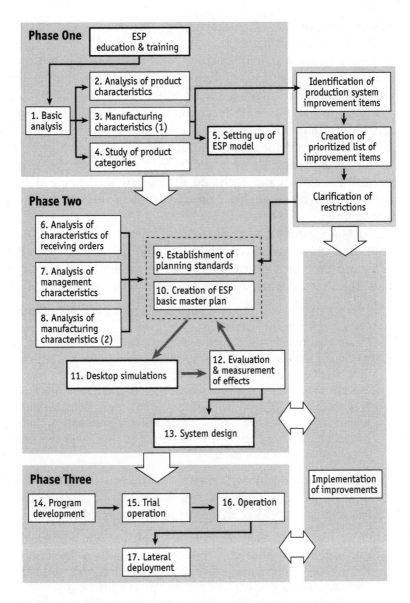

Figure 4-4. Detailed Overview of Steps for Building an ESP
Production System

inventory levels and delivery logs) that you can use later to judge
the effects of improvements. Clarifying the approach you are to take
when building equalized sizes and production patterns means
understanding the unique characteristics of your company's own
products and manufacturing system. This is so that you can take an

appropriate approach when determining matters such as equalized sizes and production patterns.

This analysis will help narrow the focus when it is time to determine the model for building a detailed system during Phase Two. The reason for having such a model is that trying from the beginning to build an ESP Production System for the entire company rather than for a more narrowly defined model area is not efficient. It would inevitably entail a lot of trial and error and wasted labor hours. Using a model is a more practical way to establish the correct approach and appropriate methods, which can later be deployed throughout the company in a much more efficient manner.

Models are established in terms of products or processes. One or the other is chosen after considering similarities among production yields and processes on the one hand or among products on the other. The following are some examples of selected models.

- When a supplier has several similar production lines, one of these lines is selected as a model.
- When a particular line is used to manufacture large amounts of a particular product (or group of products), that product (or group) is selected as a model.
- When a product is manufactured via a series of different processes that involve different production conditions, such as when upstream processes are forging processes while downstream ones are machining, you select a downstream process near the final (shipping) process as a model.

The first studies you perform to determine current conditions include three parts, 1) basic analysis, 2) analysis of product characteristics, and 3) analysis of manufacturing characteristics (1).

1. *Basic analysis.* The purpose of the basic analysis is to examine the expected effects of the new production system from the perspectives of inventory management, delivery timing, and work efficiency, and to gather data about current (preimprovement) conditions so that you will have a clear grasp of how bad those conditions are before making improvements. The targets of this analysis should be basic matters relating to production management improvement. Accordingly, each supplier may have its own matters to add to the list of analysis targets. The perspectives of the basic analysis are described further below.
   - **Analysis of inventory.** This entails learning how much inventory is being maintained for each product, which includes in-

process inventory at various manufacturing processes as well as the finished product inventory. If there are too many products to enable an analysis of each product, several representative products may be selected instead. Inventory data should include the absolute monetary value of inventory as well as its volume and value as a ratio of total sales. This is because it is hard to compare before-and-after improvements using only absolute numerical values since the total sales figure tends to fluctuate over time. Figure 4-5 shows examples of inventory analyses performed for various processes.

- **Analysis of delivery timing.** Here, surveys are done to find out how well the company has been doing in delivering orders on time. In other words, these surveys check how well or poorly the company is achieving zero missing items, which is one of the Six Guarantees of ESP. For the purposes of these surveys, any product that does not get delivered to the buyer on time is considered a missing item. In the analysis example shown in Figure 4-6, the delivery period is counted in days, but for suppliers that must deliver products at a buyer-specified hour, you should count the delivery period in hours instead.

- **Management task analysis (labor hours, steps).** This analysis investigates the steps used to carry out production management tasks and the labor hours (time) required for each step. This data will help indicate how far the supplier goes in replacing the conventional inner-circle approach to production management with ESP's logical production management approach. Progress in moving toward a more logical approach also enables you to computerize more management tasks, providing data you can use to confirm the efficiency gains of a logical production management system. Investigating steps and labor hours also helps to reveal functional weaknesses and wasteful clerical procedures, and provides information that you can use when making improvements.

2. *Analysis of product characteristics.* The purpose of this analysis is to study the relation between products or production volume trends for certain products (or product groups). The resulting data can be used to clarify distinctions among products when creating a synchronized and equalized production schedule. This involves two types of analyses.

- **Analysis of product configurations.** Here you study the production conditions (volume, etc.) for each product. Pareto charts are used to sort products by how frequently they are produced, such as in the example shown in Figure 4-7. Products are usually configured in two or three groups, with 7 or 8

Map of inventory and storage areas

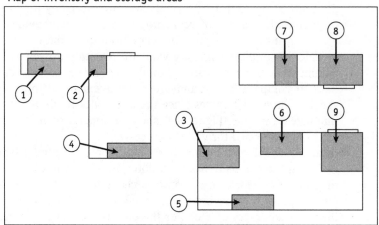

Guide to map of inventory and storage areas

Area No.	Description of inventory items	Days of inventory	Share (%) of capacity used
1	Materials	30	32.5
2	Press products (cases, parts)	30	50.3
3	Plastic parts	15	61.0
4	In-process inventory for machining	10	20.5
5	Purchased goods	15	82.3
6	Items coated by outside supplier	10	53.2
7	Products (market production)	15	43.2
8	Products (awaiting shipment)	5	22.1
9	Excess items, returned (defective) goods, etc.	20	62.2

Figure 4-5. Inventory Analysis Examples

Product A: Delivery Timing Data for September

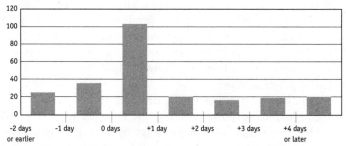

Figure 4-6. Example of Delivery Timing Analysis

levels of production volume for the various groups, and the products are then managed in two categories: those that make up a large portion of the total production volume and those that do not.

- **Analysis of products and buyers.** Pareto charts are used to represent and confirm the relationships between various products and final companies and the amounts of products to be delivered. Such analyses can be used to create patterns that confirm the effects on each final company's products. There is no need for this kind of analysis if all of the supplier company's products are being delivered to the same final company, but almost all of the supplier companies where ESP has been introduced are companies that ship products to several final companies.

Once you complete this analysis, the next step is to study how products are being delivered to your buyers.

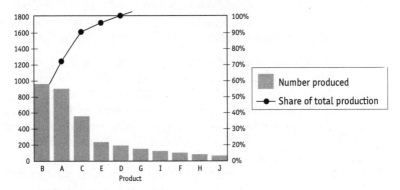

Figure 4-7. Example of Product Configuration Analysis

3. *Analysis of manufacturing characteristics.* The purpose of this analysis is to study the processes via which various products are manufactured and the lead time of each process. This is so you can clarify the bottleneck processes and problems related to manufacturing methods. This analysis requires these two steps.
   - **Step 1: Analysis of manufacturing processes and lead times.** Study the processes via which various products are manufactured and the lead time of each process. In the example in Figure 4-8, among the many products being produced only three or four major products (the ones produced in the greatest volumes) have been selected for study.
   - **Step 2: Analysis of product categories.** First you categorize products into groups according to the research done on the characteristics of products and manufacturing processes. After

that you select a model based on a consideration of how products have been grouped and whether the model is suitable for lateral deployment at a later date. When building and operating the ESP Production System, it is also a good idea to consider the product categories that will be used in the management cycle of planning, manufacturing, and evaluating results.

PHASE TWO: BUILD A SYNCHRONIZED AND EQUALIZED SCHEDULING MODEL

During Phase Two (see Figure 4-4), you use the model selected at the end of Phase One to start building a synchronized and equalized scheduling system and devising improvements in manufacturing processes. (Later in this chapter we will explain in detail synchronized and equalized planning systems and manufacturing process improvements.) The building of a synchronized and equalized scheduling model requires several in-depth analyses, which are described below.

1. *Analysis of order characteristics.* The purpose of this analysis is to determine the characteristics of orders received from buyers. This includes analyzing the distribution of order characteristics among buyers, including the variation in the amount of orders received during certain periods (monthly, weekly, daily, etc.) from the time each order is received until the ordered products are delivered, and analyzing other aspects of the amounts and timing of orders. You will use the resulting data to create production patterns that reflect the order characteristics. These analyses focus on the following three kinds of topics.
   - Topic 1: Analysis of order confirmation. Orders are received from different buyers with different timing, and it is important to analyze how well-confirmed each order is (i.e., how likely are changes after an order is received). The results of these analyses can then be applied to help determine timing and scope when planning for orders. The key question is how much time is needed between the point where a buyer's first order is received and the delivery deadline for the ordered product. When studying the way orders are processed, look at whether any changes were made in orders after they were received. If changes were made, how much time elapsed since the original orders were received (or how much time is left before the delivery deadline)? You should gather and analyze all of this data to determine how firm orders from each buyer are when they are first received.

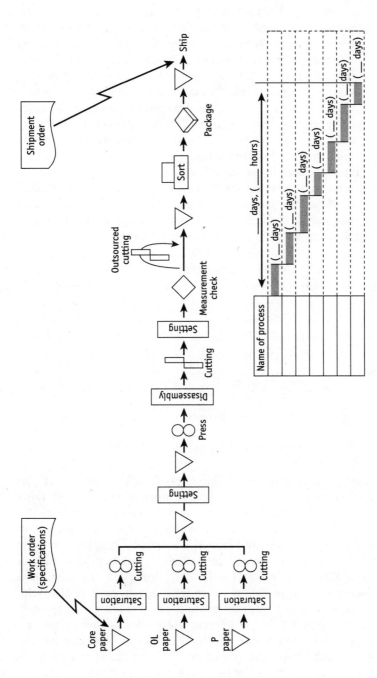

Figure 4-8. Example of Manufacturing Process and Lead-Time Analysis

- **Topic 2: Analysis of order reception timing.** This analysis includes an investigation of the timing by which order information is received from buyers and the period covered by orders. (This data is collected at the same time as the data from the order confirmation analysis.) The timing and period of order information means the cyclic periods for receiving orders, such as receiving the next month's orders on the 20th day of the previous month, or receiving the next week's orders on Thursday of the previous week. When different buyers use different ordering cycles, you must separately investigate and record each buyer's timing and period.
- **Topic 3: Analysis of variation in order amounts.** This analysis audits for periodic unevenness in delivery deadline scheduling. It may be that certain large groups of products tend to be delivered all at the end of each month or on weekends. Figure 4-9 shows weekly variation in order amounts.

Orders received (Product A)

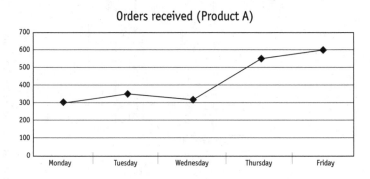

Figure 4-9. Example of Order Amount Variation Analysis

2. *Analysis of management characteristics.* The purpose of this analysis is to clarify the current production management functions and operations. When you need this sort of in-depth analysis in addition to the management analysis that was done as part of the basic analysis in Phase 1, it generally includes interviews with relevant persons in order to clarify relationships, such as those illustrated in Figure 4-10. This illustration of relationships does more than just show where relationships exist: it also helps identify problems such as missing or incomplete management functions, and it helps pinpoint the causes of these problems. Depending on the nature of those causes, when building the new production system the causes may help to clarify whether it is better to maintain certain current methods as essential to the current production management system, or whether it is better to

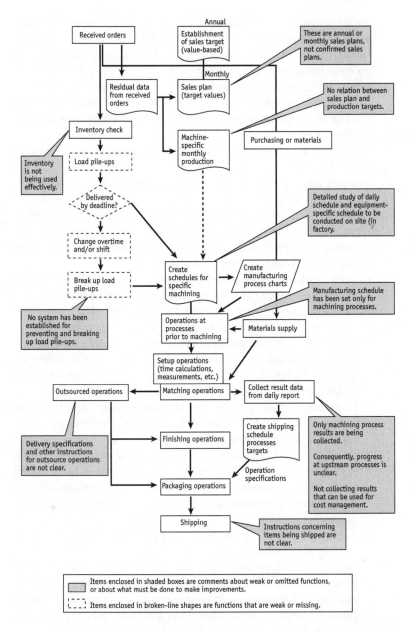

Figure 4-10. Example of Order Processing and Planning Method Analysis

make improvements in those methods. For example, if labor scheduling (scheduling how the workload will be distributed for the sake of leveled production) is currently being done based on the experience and instincts of the production-scheduling managers, this should be seen as a missing function. Looking into the causes of this problem, we may find, for example, that there are no clear standard times or standard lead times for manufacturing products. In such cases, we would add the setting of standard times and/or standard lead times for manufacturing products to the list of things to do when building the new production management system.

3. *Analysis of manufacturing characteristics.* During the basic analysis of manufacturing characteristics that was done in Phase 1, the analysis was focused on manufacturing processes and lead times. This second analysis takes a closer look at the current status of capacities (and losses) as well as purchasing operations in order to reevaluate the production framework and identify possibilities for boosting efficiency and shortening lead times. These analyses focus on the following three kinds of topics.

- Topic 1: **Analysis of process capacity.** This is an investigation into the per-hour capacity of target processes in order to clarify the balance of processes and any bottlenecks that may exist. However, since there is often a big difference between an equipment unit's designed capacity and its actual capacity, it is important to clarify the actual capacity. Usually, the difference between designed and actual capacity is due to various forms of loss, such as breakdowns and idle time during changeover. In processes that involve a lot of human labor, differences in capacity may be due to the number of people working at the processes, so you need to investigate such differences. At this point, not much attention is paid to possible differences among individual workers. That topic will be pursued later, when process improvements are being devised and implemented. Figure 4-11 shows an example of a form used to analyze the capacity of processes.

- Topic 2: **Analysis of process operation rates (load factors).** This is an investigation of the load conditions at various processes. The production load for certain combinations of products tends to become concentrated at certain processes, causing bottlenecks, so it is important to analyze load conditions. However, if production load has been rather thoroughly line-integrated so that the flow of products and the production time at each process are steady, this analysis may not be necessary.

- Topic 3: **Analysis of purchasing periods.** This includes studies of the lead time between ordering materials and delivering products, individually or in lots (see the purchasing investiga-

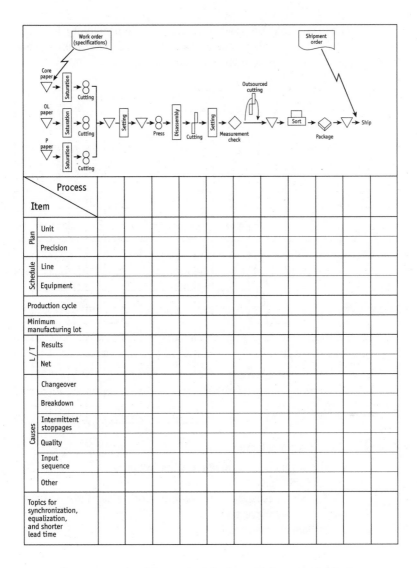

Figure 4-11. Example of Form Used for Process Capacity Analysis

tion form shown in Figure 4-12). You must take into account purchasing periods for materials when planning production schedules. When a particular type of material has a major impact on production scheduling, the lot size in which that material is purchased should be used as a criterion when establishing equalized sizes. In other words, purchasing lead-time data can be very important. When product delivery delays occur due to certain materials whose purchasing lead time is

longer than that of other materials or when delivery lot sizes
are too large, such problems can be solved by synchronizing
and equalizing purchased parts and materials.

Category	Item	Monthly usage	Purchasing unit	Amount delivered per month	Minimum delivery lot size	Delivery lead time	Delivery system			Acceptance inspection			Transaction units			Selection of vendor		Storage method	Issued by:
							Delivery method	Type of packing	Delivery unit	Inspection method	Inspection frequency	Lot size	Material costs	Machining costs	Management costs	Vendor name	Reason for selection		

Reasons for selection:

1. Has technology or equipment that we lack.
2. Less expensive than in-house.
3. Provides a capacity buffer against variable demand.
4. Avoids over-investment risk.
5. Management efficiency.
6. To build supplier relations.

Figure 4-12. Example of Form Used for Purchasing Investigations

You can use the data resulting from the various analyses performed during Phase 2 not only to establish standards for production planning but also to help determine which manufacturing
processes are most urgently in need of improvement. During Phase
2 the supplier selects a model and conducts trial operations for
building a new production system. If trial operations indicate problems, you must make improvements to complete the model.

PHASE THREE: DEVELOP AND RUN THE OVERALL SYSTEM

Phase Three (see Figure 4-4) is the phase in which you actually operate the system you built in Phase Two. It is also the phase where you determine the scope of the processing to be performed via the network host computer and PC terminals as part of the computer system, which is developed and operated in this phase. It may be necessary to reprogram some of the host computer's functions. The various items implemented during Phase Three are described below.

1. *Operation of model.* Begin trial operations of the model built in Phase 2. Deal with whatever problems arise at this trial operation stage.
2. *Program development.* This may include development of programs for the network host computer and PC terminals or, if the scope is small enough, just for the PC terminals.
3. *Manufacturing process improvements.* Preferably, you should complete all manufacturing process improvements during Phase 2. Any remaining improvement themes should be implemented starting with the highest-priority themes. It is important to enlist the cooperation of the relevant production supervisors. There are probably too many improvements to be made by production management staff alone. Although it has been stated several times already, it bears repeating that the ESP Production System cannot be built without first improving manufacturing processes.
4. *Lateral deployment.* Lateral deployment involves extending an ESP model laterally to other products and processes. It may be necessary to conduct some new analyses to address issues that differ from those in the model, but the model's standards should be applied whenever possible. The model serves as a basis for estimating the labor hours, the first processes to be addressed, the steps in building the new system for the target products and/or processes, and the schedule for building the new system.

KEY POINTS WHEN BUILDING A SYNCHRONIZED AND EQUALIZED PRODUCTION SCHEDULE

To refresh your memory, recall that ESP stands for "Equalized and Synchronized Production" and uses synchronized and equalized production methods to realize the Six Guarantees that will eliminate the pitfalls of JIT. Chapters 1 and 2 discussed the synchronized and equalized production system. (Review Figures 1-14 and 1-15 regarding the process or proposing a synchronized and equalized production schedule.) As discussed in Chapter 3, the ESP Production schedule (confirmed production schedule) utilizes ESP production patterns, making it the basic principle of developing your production scheduling.

When planning a production schedule, many companies try to plan the amount of production output based on the amount of orders received. It is a tough job for the planners, since they have to schedule their labor appropriately, so as to level out production and avoid load accumulations. The planners should keep the following points in mind when planning the production schedule.

- **Buyer's delivery deadline.** The first priority is to incorporate the buyer's delivery deadlines into the production schedule. Often, there are some buyers that submit orders with very short deadlines, and it is a real headache for schedule planners to work those tight deadlines into the schedule.
- **Prevent excess inventory.** As was mentioned in Chapter 2, planners must do everything they can to avoid creating excess inventory that ties up company assets, and this includes not only product inventory but also inventory of basic materials and in-process goods.
- **Achieve a high operation rate in terms of personnel and equipment.** A high operation rate for personnel and equipment is an obvious advantage in terms of keeping down costs. However, few planners give enough consideration to a high operation rate as a part of planning oriented toward meeting delivery deadlines.
- **Prompt responses to problems.** Production schedules are often disrupted by factory-floor problems such as equipment breakdowns or late delivery of materials to the production line. The production supervisors must be prepared to act promptly and decisively in responding to such problems. In fact, responding to factory-floor problems can also be a major worry for production planners as well. All production supervisors, managers, and planners will be much happier if they can somehow prevent such problems and keep production on schedule.

In sum, production schedule planners must meet a number of requirements while anticipating what adjustments they must make to solve problems. Having a synchronized and equalized production schedule, or as we refer to it, an ESP Production schedule, will simplify the task of making such adjustments. It will also help ensure zero missing items, prevent overproduction to keep inventory levels at their minimum, and help maximize production efficiency so that planners can meet the JIT delivery needs of various buyers. The establishment at the planning stage of equalized production volume and production sequence patterns will provide numerous advantages as the production schedule is being implemented.

The first of these advantages is that you will simplify progress management. Improvements that equalize production also improve the timing by which materials are delivered to production lines and create closer ties between processes, which results in shorter lead times. After these improvements, there is less to check on at each process and progress management therefore becomes much easier.

The second advantage of equalized production is that it simplifies volume management, such as the management of production yield. When you gear production strictly to the amount of products ordered by buyers, it is important to frequently count the volume of goods at various production stages. In particular, assembled products have many parts that must be counted unit by unit. Establishing equalized sizes, such as 10 units, and equalized units for containers as well, can greatly reduce the time required for counting parts.

When you take the synchronized and equalized production approach, there are five techniques that you incorporate in the logic that you use to plan your ESP production schedules. Figure 4-12a lists the five techniques, which we will discuss next.

Five techniques of synchronized and equalized scheduling
1. Limited inventory—separating production and buyer requirements
2. Equalized schedule—equalizing production units
3. Developing production patterns
4. Reviewing function
5. Classification of ESP production system targets—ESP highway products

---

Figure 4-12a.

Technique 1: Limited Inventory—Separating Production and Buyer Requirements

In the JIT production system, stores of in-process inventory are maintained between various processes, but in the ESP Production System the only place where you accumulate stock is at the final process in the production line. For decades, production managers have been told, inventory is evil. However, the fact is that inventory serves several useful functions, including the following.

- It **displays available items.** Like a retail store, an inventory store provides samples of available goods that buyers can directly view before making their selections.
- It **helps prevent missing items.** An inventory makes it possible to fill last-minute orders without causing missing items in other orders. As such, inventory works like a safety buffer. It enables the company to respond more flexibly to changes in existing orders or last-minute additional orders.
- It **provides a load leveling function.** When peak loads are anticipated, a supplier can make parts for certain orders in advance during low-load periods and then keep them in inventory. This type of inventory helps to reduce load variation while avoiding excess capacity (of labor and equipment).

The skilled usage of product inventory to help prevent missing items and balance the production load is part of the synchronized and equalized scheduling approach. This approach provides a solution to the problem of dealing with the delivery needs of buyers that tend to create variation in production loads. It is especially helpful in the case of suppliers that must provide wide-variety, small-lot production with frequent deliveries to numerous buyers. Suppliers that make use of the above inventory-related functions can minimize the adverse effects of such demanding orders from buyers and can also deal more effectively with the various types of loss that naturally occurs. This approach will also help suppliers meet their goals for improving production efficiency. In the ESP Production System, this type of product inventory is called limited inventory. Limited inventory can be defined as:

- Inventory that is controlled to remain at the lowest possible level.
- A buffer that helps prevent missing items, even though the supplier's production system is separate from the buyer's delivery needs.

Figure 4-13 shows a chart that illustrates variation in limited inventory in response to buyer requirements and production volume. When a buyer's requirements rise, limited inventory can help make up the difference without causing major variation in production volume.

However, one might wonder whether maintaining a product inventory would result in higher inventory levels overall. It is true that, unless you make certain improvements, inventory problems such as a lack of products or an excess of products can occur. Also, large amounts of inventory may be needed to provide an adequate

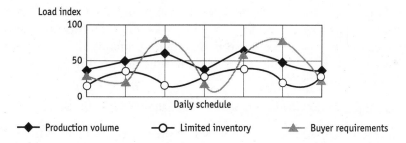

Figure 4-13. Separation of Production and Buyer Requirements

safety buffer for production variation at certain processes, and the presence of inventory between processes can create transportation problems and variation in capacity utilization. However, you can avoid these types of problems or at least minimize them by making thorough improvements in manufacturing processes so that product inventory does not become dead (useless) inventory or long-term (costly) inventory. When you make such improvements, you can reduce overall inventory levels by 20 to 50 percent even while maintaining limited inventory levels.

One key question is "How many locations of limited inventory are needed?" The answer to that question depends on the supplier's order characteristics and manufacturing characteristics, and therefore you must consider the results of the analyses of those characteristics when determining the amount and locations of limited inventory. The following is an example of how to set the number of locations of limited inventory from a supplier that has adopted this system.

- **Method used to set standards for equalized sizes.** When production is based on equalized sizes, you can set up multiples of the equalized size as the limited inventory, based on a consideration of the order variation data. However, the inventory levels must never exceed the current level.
- **Method for setting targets.** Limited inventory levels are established as levels that are a certain percentage less than the current product inventory levels. This may not seem very logical, but we must remember that when building ESP it's assumed that the supplier will make various improvements, and therefore you must set target values that reflect the expected postimprovement conditions. Your manufacturing processes are improved and the new production system is built with a view toward achieving these targets.

Whatever limited inventory levels you initially set should not be regarded as cast in stone. As you make manufacturing process improvements, you should reevaluate and gradually reduce the target limited inventory levels as the improvements allow.

Technique 2: Equalized Schedule—Equalizing Production Units

An equalized schedule is one in which the target products of the ESP Production System are always manufactured in the same amounts (an integral multiple of the manufacturing lot size). Likewise, the parts and materials for these products are always purchased in the same (equalized) amounts (see Figure 4-14). This differs fundamentally from the JIT approach of only what is needed and only when it is needed. A production system that turns out more than just what is needed and when it is needed may seem disadvantageous at first when compared to JIT, but this difference is made up for since there is no need for time-consuming management work to balance JIT production volume with production staffing, and it tends to cause much less confusion on the factory floor.

Production schedule based on equalized sizes

Item number	1	2	3	4	(5)	(6)	7	8	9	
A	50						50			
B		20						20		
C			30						30	
D				80						

(Equalized sizes are used in all production stops and insertions.)

Figure 4-14. Image of Production Schedule Based on Equalized Sizes

The method used to determine equalized sizes depends on various factors. Figure 4-15 shows a decision-making flow chart that you can use to determine a suitable method. This flow chart represents a general approach. In practice, this decision should also be based on the results of various analyses. In other words, it should be based on the specific characteristics of the company in question. Therefore, the decision-making flow chart in Figure 4-15 should be considered as only a general reference for determining equalized sizes. We will now look at some examples for determining equalized size (Figures 4-15a to 4-15e).

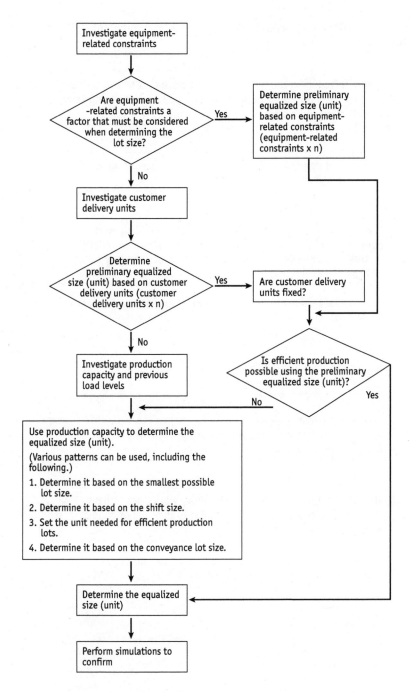

Figure 4-15. Approach for Determining Equalized Sizes

Example 1: Determine equalized size based on equipment-related constraints.

Before introducing the new production system, sake was bottled based strictly on the amounts to be shipped, so the mixing tank levels had to be constantly adjusted accordingly.

A equalized size for bottling was determined according to the tanks' capacity.

The results were that the tank levels did not need so much adjusting and the steady bottling lot size meant that ordering of materials such as cartons and management of inventory became simpler and easier.

Figure 4-15a. Determining Equalized Sizes for Mixing Tanks at a Sake Brewery

Among the various processes, constraints were determined to exist at the heat treatment process.

An equalized size was determined for the amount that can be produced by the heat treatment process.

The labor hours required for preparatory work prior to heat treatment, such as setting up for various amounts of goods, were greatly reduced.

Figure 4-15b. Determining Equalized Size According to Heat Treatment Process Constraints at a Foundry Company

Example 2: Determine equalized size based on delivery units.

Deliveries were made every day, but products were manufactured only in monthly batches.

An equalized size was determined as a multiple of the amount delivered daily.

- It greatly reduced the labor hours required to check quantity before shipping.

- At the same time, the equipment used to convey in-process inventory was improved to handle equalized units, so fewer labor hours are required to check the quantity of in-process inventory.

Figure 4-15c. Determining Equalized Size as a Multiple of the Units Delivered by an Automotive Parts Manufacturer

Example 3: Determine equalized size based on smallest possible lot size.

The production volume and timing are vague for each production run. There is only one production run per month.

The elements defined below were used to calculate an equalized size:

- Available production time per month: Pt (Production time)

- Overall average load time per month: L (Load)

- Total production volume per month: Qp (Quantity of production)

- Available changeover time per month: Ct (Changeover time)

- Changeover time per production run: C (Changeover)

- Number of changeovers per month: Tc (Times for changeover)

- Equalized size: E (Equalized size)

Pt (available production time per month) − L (overall average load time per month)
   = Ct (available changeover time per month)

Ct (available changeover time per month) / C (changeover time per production run)
   = Tc (number of changeovers per month)

Qp (total production volume per month / Tc (number of changeovers per month
   = E (Equalized size)

Example 3 *(continued)*

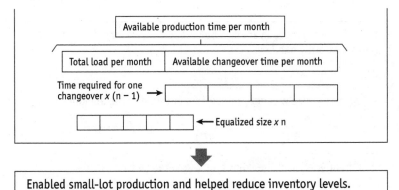

Enabled small-lot production and helped reduce inventory levels.

Figure 4-15d. Determining Lot Size for a Pharmaceutical Manufacturer's Packaging Process

Manufacturing plant included over 40 die-casting machines and 10 processing lines.
- Monthly unit was set for each equipment unit and each line based on the order information.
- Die changover was not done until the specified production volume was completed, regardless of the time of day.
- Operations were always subject to whatever changes or additions buyers wished to make in the order information, and managers had to frequently check on progress and issue modified specifications to the production staff.

It is more efficient to change dies at the beginning or end of the shift, considering the need for die maintenance.
- The production volume per shift is set as the equalized size.

It is more efficient to change dies at the beginning or end of the shift, considering the need for die maintenance.

Since the changeover times have been fixed, it becomes possible to shorten the setup time.
- Labor hours spent in planning have been reduced since there is no longer any need for complicated volume calculations as part of planning.

Figure 4-15e. Determining Equalized Size per Shift at an Aluminum Cast Manufacturer

Example 4: Determine equalized size based on shift unit.

- This aluminum cast product manufacturer used a multiple of the amount that can be produced per shift (1 shift = 8 hours) to equalize lot size.
- From 30 minutes to one hour had been needed for die maintenance at the end of each shift. The company planned to improve efficiency by using that time for die replacement.

Technique 3: Developing Production Patterns

Developing production patterns includes confirming which products you are to input to which line and determining in advance an efficient input sequence. In other words, the production amount for each item is set according to recent sales results or sales projections. Then you use these production amounts as a reference when planning a production schedule that includes an input sequence and production lot size that are geared mainly toward maximizing production efficiency.

The managers who plan production schedules tend to give utmost priority to the deadlines and other requirements and opinions of buyers and sales staff. As a result, they tend to plan production input sequences accordingly, with little regard for production efficiency, and this can cause dissatisfaction among the factory-floor staff. If they can manage to keep production efficiency in mind when planning input sequences, they will naturally (and perhaps even unconsciously) produce a production schedule that has better production efficiency. This approach not only tends to reduce the labor hours needed to plan a production schedule, it also makes planning tasks easier to computerize.

ESP production patterns (based on production patterns) must be revised as needed when factors such as market demand and preliminary in-house information change. When production pattern techniques are applied in businesses such as the process industry, it becomes possible at the planning stage to avoid loss due to model changeover time. In fact, the input sequence can have a major impact on product quality as well as productivity. There are two methods for establishing ESP production patterns, which differ depending on the type of analyses being done.

The first method starts by studying the results of past shipments to get an idea of the supply and demand levels. This enables the most efficient input sequence and production volume to be determined in advance, as illustrated in the production pattern shown in Figure 4-16. This method has already been described in Chapter 3, so it will not be described here.

Production schedule based on equalized sizes

Item number	1	2	3	4	⑤	⑥	7	8	9	10
A	50						50			
B		20						20		
C			30						30	
D				80						80

The input sequence that provides the maximum production efficiency (in this case, A ➡ B ➡ C ➡ D➤ becomes the production pattern.

Figure 4-16. Image of Production Pattern

The second method is used in cases where too much variation in past results or in demand predictions makes it impossible to create a single pattern that can be used for each production run. In such cases, there are several input patterns, each based on a different set of parameters, such as changeover time and number of production runs. This second method is illustrated in Figure 4-17. For efficiency, the most important factor is changeover time. When products are listed in a matrix such as shown in Figure 4-17, it becomes easier to check different product combinations and identify which combination has the shortest changeover time.

The number of production runs is determined based on the equalized size (or a multiple of the equalized size) of the orders for a certain period. Once you have determined the number of production runs, an input pattern can be created according to the number of production runs, as shown in the figure, and you can then implement production according to that pattern. When creating input patterns based on the number of production runs, it is important to keep efficiency in mind when arranging combinations of products. In the example in Figure 4-17, the patterns are based on orders that come in monthly units.

If it is not possible to determine the optimal production sequence in advance, create a product changeover matrix and then create input patterns based on the number of production runs, or create production patterns based on information such as sales plans.

Input patterns based on number of production runs

Product changeover matrix

	A	B	C	D	E
A		10	30	35	20
B	10		40	20	30
C	30	40		35	45
D	35	20	35		25
E	20	30	45	25	

	No. of runs	Week 1	Week 2	Week 3	Week 4
A	Two runs	●		●	
	Three runs	●	●	●	
	Four runs	●	●	●	●
	Six runs	●●	●	●●	●
B	One run		●		
	Two runs		●		●
	Three runs	●	●	●	

Determine the number of production runs based on information such as sales plans and determine the input sequence based on a product changeover matrix.

Figure 4-17. Group-based Production Sequences

Planning method before improvement

- Monthly schedules were planned based on monthly sales information.

- Shipments tended to increase during the winter cold and flu season. The company responded to this trend by front-loading the products for winter into production volumes for August to December.

Results of investigations and new method for creating pattern

- Investigation of previous monthly shipment data for the target products shows little variation from year to year.

- The front-loaded production volume for winter did not show much variation. It was determined that inventory imbalances could be avoided simply by producing more starting in January.

- The company determined new production volumes for various products and for each month of the year.

- They decided to use two types of monthly production sequences (patterns):

  Pattern for cold and flu medicines.
  Pattern for other medicines.

Results of improvement

- Schedule planning labor hours were reduced 30 percent.

- Product inventory levels were also reduced 30 percent.

Figure 4-17a. Example 1: Production Pattern Based on Shipping Data at a Pharmaceutical Company

Although this example uses individual products, you can just as easily use product groups. Grouping works well if there are several similar products and you wouldn't need much time to switch among the products. Grouping is less effective if the products are too diverse and require long changeover times. The following two examples (Figures 4-17a and 4-17b) show how two companies determined their production patterns.

Conditions before improvement

- Production periods and volumes were determined based on orders from buyers.
- Production was planned for one month's production volume.

Results of investigations and new method for creating patterns

- Customers included almost 10 different end-buyers, whose volume of orders varied a lot from month to month and showed no consistent trends.
  - Changeover times for each specific buyer were only about 30 minutes, since the products for each buyer were similar in shape. However, when stitching to products for a different buyer, changover time ranged from three to six hours.
  - Products were grouped according to the buyer, changeover times between groups were clarified, and the production sequence was arranged to obtain the shortest possible changeover times between groups.
  - Improvements were made to reduce changeover times and thereby enable small-lot production. These improvements enabled the company to stop planning production in monthly units and instead switch to several production runs per month.
  - The company was able to establish a monthly production schedule using input patterns for each number of production runs and using the changeover matrix.

Figure 4-17b. Example 2: Production of Pattern Based on Changeover at an Automobile Parts Manufacturer

Technique 4: Reviewing Function

Performing reviews under ESP becomes easier because of the use of patterns in production scheduling and because you can add or subtract equalized sizes to adjust the production load. Once you have incorporated received order information (preliminary information or confirmed information) from your buyers into your production patterns, you then conduct a review, which also includes the use of simulations, on each of the following.

1. Zero missing items
2. Overproduction prevention
3. Load feed forward

During each review you use ESP's stops and insertions to make adjustments in line with the review's objective (see Figure 4-18). These three reviews functions are described in more detail below.

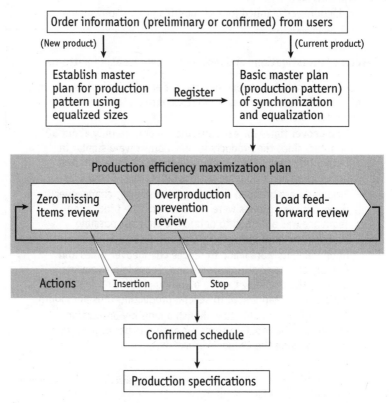

Figure 4-18. Conceptual Diagram of Simulation

Zero missing items review

This review checks for the occurrence of missing items in production based on production patterns that incorporate preliminary or confirmed information from buyers. If the production patterns were created without first incorporating preliminary or confirmed information from buyers, then you must use requirement information from buyers to check for missing items in order to ensure zero missing items.

Whenever you find missing items has occurred, you perform an insertion using an equalized size. You perform this check for missing items just before issuing the production specifications. Figure 4-19 shows an example of actions you take when you find missing items during this review.

Basic data for creating a production pattern
• Limited inventory: 100
• Equalized size: 50

Information from zero missing items review
Current inventory: 70

Item number	1	2	3	4	5	6	7
Production	50			50			50
Preliminary or confirmed information	30	30	30	50	20	20	20
Inventory trend	90	60	30	30	10	-10	20

⬇   (Missing items)

Item number	1	2	3	4	5	6	7
Production	50			50		50	50
Preliminary or confirmed information	30	50	50	50	20	20	20
Inventory trend	130	80	30	30	10	40	20

(Insertion)

Figure 4-19. Model of Zero Missing Items Review

Overproduction prevention review

This review looks for inventory levels that exceed the limited inventory level in the production schedule based on a production pattern in which you have incorporated preliminary or confirmed order information from buyers. You insert a stop to cancel production of an equalized size in order to prevent overproduction. The timing of the overproduction prevention review is the same as that of the zero missing items review. Figure 4-20 shows an example of the actions you take when you find overproduction during this review.

Load feed-forward review

This review incorporates preliminary or confirmed information from the week (or 10-day period) following the week (or 10-day period) that was reviewed by the first two reviews. If the production volume required by buyers for that period is greater than the production volume in the previously created production pattern, or if the load has increased after performing the missing items or

Basic data for creating a production pattern
• Limited inventory: 100
• Equalized size: 50

Information from overproduction prevention review
Current inventory: 20

Item number	1	2	3	4	5	6	7
Production	50			50			50
Preliminary or confirmed information	10	0	0	10	15	15	15
Inventory trend	60	60	60	110	95	80	80

⬇ (Overproduction)

Since the inventory exceeds the limited inventory level on the fourth day, a stop is inserted during that day

Item number	1	2	3	4	5	6	7
Production	50			0			50
Preliminary or confirmed information	10	0	0	0	15	15	15
Inventory trend	60	60	60	60	45	30	65

(Stop)

Figure 4-20. Model of Overproduction Prevention Review

overproduction prevention review, you feed some of the production volume forward to level out the production load (see Figure 4-21).

Technique 5: Classification of ESP Targets—ESP Highway Products

As was explained earlier when describing ESP's separation of production and buyer requirements, the ESP Production System includes limited inventory (final product inventory). However, you do not include all types of products you're manufacturing in this limited inventory. A company would expose itself to considerable loss if it included products in limited inventory that were only manufactured in small lots, or products that it was not certain would be ordered anytime soon. Likewise, there is no need to include in the ESP production pattern any products that will not be ordered soon. Figure 4-22 illustrates the classification that is used to determine which products are target products for limited inventory and production patterns.

Basic data for creating a production pattern
- Limited inventory: 100
- Equalized size: 50

Information from review
Current inventory: 60
Preliminary or confirmed information from following week:
  160 (55 more than the predicted amount)

Item number	1	2	3	4	5	6	7	8	9	10	11	12	13	14
Production	50			50			50				50			50
Preliminary or confirmed information	15	15	15	15	15	15	15	65	20	15	15	15	15	15
Inventory trend	95	80	65	100	85	70	105	40	20	5	40	25	10	45

Insertion

Item number	1	2	3	4	5	6	7	8	9	10	11	12	13	14
Production	50			50		50	50				50			50
Preliminary or confirmed information	15	15	15	15	15	15	15	65	20	15	15	15	15	15
Inventory trend	95	80	65	100	85	120	155	90	70	55	90	75	60	95

The production volume is 55 more than the amount predicted when
the production pattern was created, so an insertion is made on one
of the days in the following week (the second, third, or sixth day).

Figure 4-21. Example of Load Feed-Forward Review

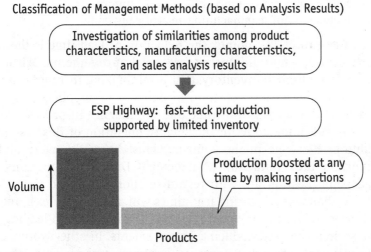

Classification of Management Methods (based on Analysis Results)

Investigation of similarities among product
characteristics, manufacturing characteristics,
and sales analysis results

ESP Highway: fast-track production
supported by limited inventory

Production boosted at any
time by making insertions

Volume

Products

Figure 4-22. Classification of Management Methods

Basically, this classification looks at product characteristics, sales characteristics, and manufacturing characteristics. Once you have analyzed the similarities among these characteristics (such as similarities among products), then you can further clarify and classify frequently repeated orders, product delivery records, production volume, and production efficiency. The products that are designated as targets for limited inventory are given priority as fast-track products in terms of ensuring prompt delivery, and so we call this group of products ESP highway products.

KEY POINTS FOR ESP PRODUCTION SYSTEM INNOVATIONS

In the context of the ESP Production System, manufacturing process improvements are factory-floor innovations and improvements that are necessary to build and operate ESP and to ensure its growth and success. These innovations and improvements are targeted at the following three areas.

1. Synchronization to improve delivery to buyers
2. In-house manufacturing system
3. Outsourced goods and purchased parts

The improvements to be made in these three areas will be described after we consider three key points for manufacturing process improvements.

1. Reduction of lead time
2. Minimization of in-process (limited) inventory
3. Elimination of manufacturing-related constraints

Of these three key points, the reduction of lead time is the most important for manufacturing process improvements. Although maintaining limited inventory, that is minimizing in-process inventory, to enable synchronization of delivery to suppliers is a basic part of ESP, reducing lead time and minimizing in-process inventory are still imperative for the smooth operation of this system. In addition, everyone in the company must be made aware of how important the Production Management Division's functions and leadership position are with regard to shortening lead time. After all, the challenge of cutting lead times will, at the very least, require that production management staff take the initiative in leading and overseeing lead-time-reducing improvements. In other words, these improvements are not incremental improvements made via the traditional bottom-up approach. Instead, you must coordinate them from the top down, as improvements aimed at overall optimization of the production system. This means that it is the Production Management Division's role to develop improvement themes in line with the objectives of the ESP Production System and to vigorously promote implementing these improvements to reduce lead time.

The maintenance of a limited (minimization of in-process) inventory for principal products enables factory-floor staff to avoid being directly impacted by variation in deadline schedules and production volumes. As was mentioned earlier, many companies who have a typical level of manufacturing flexibility are faced with a trade-off between ensuring prompt delivery (i.e., providing good service to buyers) on the one hand, and ensuring high productivity in their manufacturing activities on the other hand, and typically these companies give service to buyers priority over their own productivity in this trade-off. However, this approach is putting the squeeze on companies as they also face the need to cut costs in order to survive in today's economy and marketplace. Consequently, boosting productivity is simply a must in order to help keep costs down relative to output. But there is an approach by which companies can turn this zero-sum (trade-off) situation into a win-win situation—separating production from buyer requirements. This brings us back to the three areas where you need to target your innovations and improvements.

Target 1: Synchronization to Improve Delivery to Buyers

Synchronization with buyer delivery deadlines is the first area where you need to target your innovations and improvements. This means maintaining a limited inventory of principal products so that the company can:

- Have a quick responsive production system.
- Ensure zero missing items.
- Use synchronization as a means of maximizing production efficiency.

You need to always keep in mind that the ESP Production System is not something that is established, approved, and then left alone. Rather, it requires ongoing improvement and fine-tuning. Likewise, the number of products being kept in limited inventory does not stay the same year after year. Besides, your eventual goal is to achieve zero limited inventory.

Whatever limited inventory you have in the factory, you must manage it using visual management techniques. The amount of limited inventory is a factor that you can use to visually confirm the results of improvements (specifically, structural improvement of the production system). Therefore, visual management is a must. It is also important to make sure you are managing all inventory in as simple a way as possible and that all your operations are in compliance with the relevant standards and rules.

Target 2: In-house Manufacturing System—Synchronization and Equalization of Manufacturing Processes.

The second area where you need to target your innovations and improvements is with your in-house-manufacturing processes. Actually, under the ESP approach, activities that help minimize manufacturing lead times are also activities that help synchronize and equalize manufacturing processes. The ultimate goal is to ensure synchronization with buyer deadlines combined with reduction of the limited inventory level to zero. Generally, a large percentage of manufacturing lead time consists of idle time for lots undergoing machining or other processing, standby time while awaiting conveyance between processes, or bottlenecks at outsourced supplier factories. Companies should therefore strive to eliminate retention of goods at all manufacturing processes, from the first one to the final one. Figure 4-23 lists some common perspectives concerning the kinds of improvements you need to make in your in-house manufacturing processes.

Analyses of your manufacturing characteristics, as well as other analyses, can help you identify the best approach to devising improvements for those issues that promise to do most to eliminate problems and/or improve production planning. You should therefore resist the temptation to start by addressing whatever problem is noticed first when studying current conditions at factories. Instead, look for the problems whose elimination is most likely to shorten manufacturing lead time. Before devising improvements for those problems, be sure to perform a cost-benefit analysis. Ordinarily, many of the improvements made by production division staff are for the purpose of boosting productivity or lowering costs, and seldom are they tied in any way to improving production planning processes. Even rarer is the factory that requests improvements in production planning.

Perspective 1: Simplify processes
- Innovations in manufacturing methods
- Integration and combination of processes

Perspective 2: Regulate and line-integrate manufacturing
- Line integration of equipment layout
- Synchronization of production lines
- Line integration of isolated processes
- Changeover improvements to enable production of smaller lots
- Bring outsourced processes in-house
- Review and optimize staff assignments

Perspective 3: Elimination of bottleneck processes
- Accelerate operation times at processes
- Thoroughly eliminate loss
- Review and improve equipment use methods and operator work methods

Perspective 4: Minimize equipment-related loss
- Eliminate the "six losses"

Figure 4-23. Perspectives on Improvements to be Made in In-house Manufacturing Processes

The ESP Production System is therefore most unusual in its emphasis on making production planning-related improvements. We will now discuss the common perspectives for improving your in-house manufacturing process listed in Figure 4-23.

1. *Simplify processes.* The two areas to look into in order to simplify processes are 1) innovations in manufacturing methods, and 2) integration and combination of processes. Both of these areas require some technical skills and many proposals are likely to emerge from a review of existing processes. If you have used the same manufacturing methods for many years, there are probably numerous blind spots, and the factory managers may not be aware of the most recent and advanced manufacturing methods. You should gather and analyze all of these types of information before brainstorming for improvement ideas. Even after various improvements have been proposed, you will need further analyses to determine their feasibility in terms of cost and technology.

2. *Regulate and line-integrate manufacturing.* A close look at how parts and materials move between equipment units and between processes will reveal many instances where goods are being retained as in-process inventory. A lengthy study of various factories will

no doubt turn up some of the following reasons why goods are retained between processes.

- When equipment units are arranged by function (i.e., several similar machines are lined up together), a wide variety of products become concentrated in that area as retained goods, waiting to be processed.
- The next process or the next machine is far enough away that individual products are retained until they form a group large enough to be transported as a batch.
- Products are being processed in large lots, so there is a lot of standby time for individual products before and after being processed.
- The specifications issued on the manufacture date are not clear, so products are left idle while clarification is sought.
- Products are retained while awaiting results from between-process inspections.
- Overproduction has occurred at an upstream process, resulting in more standby time (waiting for processing) at downstream processes.
- The production line is idle due to late delivery of subassembly parts or parts from an outside supplier.

The following are some countermeasures that you can consider as ways to eliminate retention of goods due to the causes listed above.

- Line integration of equipment layout.
- Synchronization of production lines.
- Line integration of isolated processes
- Changeover improvements to enable production of smaller lots.
- Bring outsourced processes in-house.
- Review and optimize staff assignments.

At processes where equipment is arranged by function (where machines with similar functions are grouped), line integration should be implemented whenever possible. Although it is usually impossible to set up a specialized line for each type of product, grouping together products that have similar sets of processes can often make some degree of line integration possible. One simple technique is to integrate processes that use only small equipment units. Even in cases where some processes have been outsourced, line integration should be pursued if it is evident that it would result in fewer machines and/or operators. However, whatever method you use to integrate the processes it must be studied with regard to its impact on lead time. Another prime target for potential improvements is any process that involves long changeover times. When changeover times are shortened per lot, standby time between processes shrinks and the flow of goods through processes speeds up. Thus, there are many ways in which improvements can be made, and it pays to go after them aggressively. We will continue the discussion of perspectives in Figure 4-23.

3. *Elimination of bottleneck processes.* When one or more processes require an inordinate number of labor hours, it can become a major obstacle for the entire flow of goods in production. Most manufacturers are aware of the problem of bottleneck processes and have worked hard to eliminate them. However, this cannot be a one-time battle. Companies must remain ever vigilant against bottleneck processes, since they can easily turn up whenever new product types or manufacturing equipment models are introduced.

4. *Minimize equipment-related loss.* If the workplace uses a large amount of machinery, the first priority should be to thoroughly eliminate equipment-related loss. No amount of line simplification and regulation will keep production on schedule if the equipment frequently breaks down, operates intermittently, or otherwise requires frequent maintenance. Figure 4-24 describes six types of equipment-related loss, all of which should be thoroughly eliminated by implementing manufacturing process improvements.

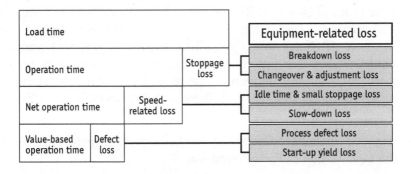

Figure 4-24. Six Types of Equipment-related Loss

Measurements of loss at companies that have not implemented improvements to eliminate these types of loss have turned up loss rates as high as 40 to 50 percent. At these companies, employees in production are resigned to the fact that their equipment will

occasionally break down or operate intermittently. There is no reason to give up like that—with the right approach and techniques you
can greatly reduce equipment-related loss. The key is to take action
to make improvements, without giving up. The *equipment's total
efficiency rate* is a value used to calculate equipment-related loss.
Figure 4-25 lists some loss calculation formulas that use this value.
The equipment's total efficiency rate can be very helpful in clarifying and analyzing sources of loss.

Among the various types of improvements you need to make,
implementing changeover-related improvements should be a high-
priority item. Changeover times will never become shorter if the
production managers believe that changeover is necessarily time-
consuming or that changeover time is not a form of loss. All
changeover time must be recognized as loss, and you must minimize this type of loss in every way possible. As members of a consulting team, the authors helped to lead a changeover loss reduction
campaign that reduced one company's changeover loss time of
four to six hours down to about 10 minutes after improvement.
Figure 4-26 outlines steps that you can use to make changeover
improvements.

In addition, there is the issue of personnel-related loss. You must
also address any personnel-related bottlenecks with loss-eliminat-
ing improvements. Personnel-related loss includes loss due to inefficient work methods and performance-related loss. To eliminate
these types of loss, closely study how people work and try out new
work methods, new jigs and tools, or new operation sequences to
improve efficiency and operability. Eliminate wasteful motions
wherever they are found. Also, you can eliminate variation (in
speed, etc.) among individual operators by establishing standard
work procedures and teaching them to every operator.

Target 3: Outsourced Goods and Purchased Parts—Synchronization and Equalization of Parts Purchasing

Outsourced goods and purchased parts is the third area where you
need to target your innovations and improvements. The purchasing
of parts and materials is greatly affected by the precision of production planning, and many companies end up trying to ensure an
ample supply of parts and materials by flooding the line with them
every day. Even though they recognize this as a problem, companies are primarily concerned with meeting their buyers' orders and

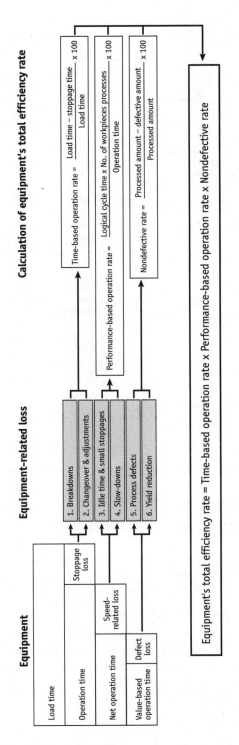

Figure 4-25. Formulas Using the Equipment's Total Efficiency Rate

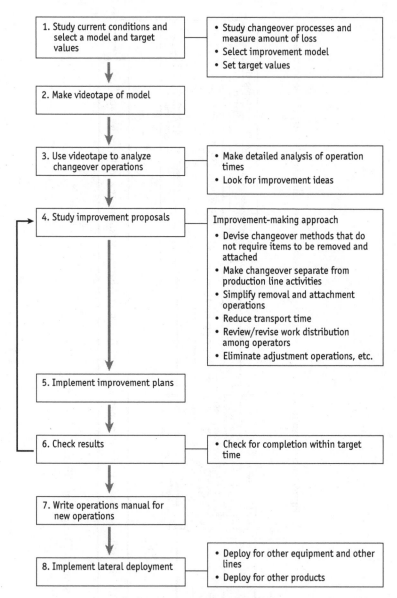

Figure 4-26. Changeover Improvement Steps

so they fail to devise any serious countermeasures. Some of the main reasons for inefficient purchasing functions are:

- Line stoppage due to defective items.
- Obsolete or stale parts and materials.
- Inventory load of parts and materials.

No matter how efficient the production sequence planned via synchronized and equalized scheduling, all your planning will be for naught if a shortage of parts and/or materials delays the schedule. Problems that generally occur in purchasing functions include the following.

- The Production Management Division and the Purchasing Division for outsourced parts and materials are isolated from each other and thus have never tried to coordinate their improvement efforts.
- Computerization of the purchasing (MRP) system has created a complicated system geared toward purchasing just what is needed, and this system is not very adaptable to schedule changes.
- Purchasing functions are not promptly updated in response to the product changes or revisions that occur almost daily.
- The production planning system emphasizes production and is usually not comprehensive enough to fully incorporate purchasing functions.

Under ESP, the production management system is evaluated with a view toward making concurrent improvements in purchasing that take the above issues into consideration and that extend the concepts of synchronization and equalization to purchasing, as discussed below.

1. *Review of purchasing methods when planning improvements in scheduling.* As was mentioned earlier, few companies concurrently pursue improvements in purchasing while seeking to improve their production systems. When building ESP, companies make the Purchasing Division a member of the project, such as by establishing a team of purchasing improvement leaders. When the production system uses equalized sizes and works more efficiently, the Purchasing Division will enjoy benefits such as the elimination of last-minute delivery order changes and the establishment of steady delivery amounts. In other words, when the people who order parts and materials have created a steadier flow of orders, the people who receive those parts and materials orders can have a steadier flow of operations. The end result is a more precise system of ordering and purchasing parts and materials. In addition, when the purchasing staff has a steadier flow of operations, they are better able to negotiate with suppliers to shorten lead times (see Figure 4-27).

2. *Revision of purchasing to reflect improvements in ordering methods.* Although many companies use an MRP system for purchasing, such systems generally fail to provide adequate functions for responding to changes in orders, in product models, or other

---

Improvement concepts for purchasing functions

1. Implement improvements related to synchronization and establishing a limited inventory; increase the precision of production planning and reflect these improvements in the purchasing system.

2. Incorporate the concept of equalization to simplify volume management and to make purchasing tasks less cumbersome.

3. Categorize the use characteristics of various parts and establish synchronization levels.

Figure 4-27. Improvement Purchasing Functions

purchasing-related factors. Conditions behind the inability of companies to respond quickly to change include the following.

- Such changes are frequent and they tend to have lead times that are shorter than the established lead time, which makes it difficult for companies to deliver on time.
- The inventory records stored on computers do not always match the actual inventory levels, so parts shortages or confusion concerning supplied parts can easily arise.
- Master records are not updated promptly enough, so sometimes the wrong parts are delivered.

Such conditions obviously indicate that these companies need to create and thoroughly enforce new rules concerning inventory management and maintenance. However, many companies have had trouble recognizing and resolving even such obvious issues as these. In many cases, they have neglected to teach their staff the importance of addressing these types of conditions, or the top managers are simply unaware of them. For these companies, it is time for everyone—including top managers—to go back to square one and start implementing the following sorts of activities.

- **Make workplaces more organized and orderly to facilitate volume management of goods.** These improvements will make it possible to understand, at a glance, which parts are where and set up ways to make numbers of parts easy to track.
- **Thoroughly reevaluate the rules.** Set up inventory management rules and master records maintenance rules and train the relevant staff members to follow them strictly.

If necessary, the company should also reevaluate their ordering system. Basically, there are two types of ordering methods.

1. *Irregular ordering of regular amounts.* This method is also known
    as the *order-point method,* by which orders are issued automati-
    cally whenever the inventory level shrinks to a certain prespeci-
    fied level. This method has the advantage of sparing the
    purchasing staff a lot of management work, and they do not need
    to worry about inadvertently omitted orders. This method is often
    used for inexpensive inventory items.
2. *Regular ordering of irregular amounts.* Under this method, the
    employees in purchasing check the inventory levels at regular
    intervals and then determine which parts and/or materials need to
    be ordered. This method enables much better control of inventory
    levels. This method is often used for expensive inventory items or
    items ordered in large quantities.

When you manage all parts and materials so they remain at the
same level, a lot of management labor is required and missing items
tend to occur more easily. However, when a company uses equalized
sizes it can switch to the order-point method and eliminate many
management tasks while also helping to minimize missing items.

Reevaluation of Parts and Materials Purchasing Starting at the Product Design Stage

The previous section explained how you could improve purchasing
operations by increasing the planning precision of the ordering system and by using equalization to simplify volume management.
This section will explain how you can improve specifications at the
product design stage to help standardize parts and materials purchasing, increase the number (i.e., reduce the variety) of common
parts and materials, and make them easier to manufacture.

The trend toward wide-variety, small-lot production in the context of rapidly diversifying and distinctive market needs is unavoid-
able nowadays. However, growth in the variety of parts and
materials places a great burden not only on the purchasing operations of companies but also on the management and manufacturing
operations of the suppliers who serve the purchasers, especially as
companies attempt to ensure a synchronized supply system for parts
and materials. The greater the variety of items, the more you must
do to keep track of them. Likewise, as processes (machining, etc.)
become more diverse, changeovers tend to get more complicated
and time-consuming, so that manufacturers must respond by
assigning more and more people to their production lines. This

pushes up manufacturing costs, as do the greater amounts of parts and materials (similar amounts per product model, but for more models) in inventory, which also entails a higher risk of having obsolete or stale inventory.

Manufacturing companies are seeking to further standardize and simplify their product lineups for the sake of higher productivity, even as they seek to expand sales by meeting the market's needs for greater product diversification and distinctiveness (see Figure 4-28). The following are some key considerations that companies should make when dealing with the seemingly contradictory needs for standardization on the one hand and diversification on the other.

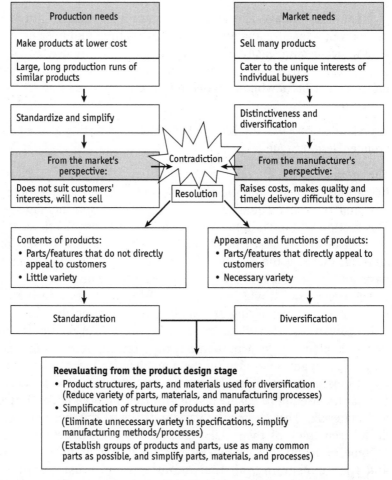

Figure 4-28. Responding to Diversification from the Product Design Stage

- Allow for diversification (in product appearance, functions, performance features, etc.), but do everything possible to minimize the amount of diversification in in-house operations.
- The causes for diversification due to a company's in-house operations are many. They might include technical innovations that have appeared since the company's products were first developed, changes in the approach taken by product designers, or changes in manufacturing equipment and technologies since the time the company's own manufacturing process was designed and built.
- Distinguish between basically necessary product diversification and unnecessary diversification, and then try to eliminate product variety that is not directly tied to buyer needs.
- Respond to product diversification by optimizing the overall product lineup based on examinations of diversification of products and parts in terms of the specific function and cost of each part.

In sum, the best approach is to organize products and parts into groups and to promote standardization of parts, materials, and processes based on common components or characteristics. The next step is to work from the product design stage to devise new product structures and parts and materials specifications that support diversification while enabling a reduction in the number of parts and materials. You can also use design work to reduce variety among manufacturing processes, eliminate unnecessary diversity in specifications, and promote simpler manufacturing. In other words, all products, parts, and manufacturing processes should be redesigned for simplicity and for ease of manufacturing while also enabling a sufficient range of diversity.

There are five concepts behind the assorted techniques used to review product designs: fixed and variable, modularization, multifunction design, range extension, and systematization (see Figure 4-29), described briefly as follows.

1. *Fixed and variable.* You should perform the following types of investigations to minimize the varieties of parts, materials, process routes, equipment, jigs, and tools, even while supporting a wider variety of product models and specifications.
   - Put all product structures, parts, materials, etc. into one of two categories: fixed components or variable components.
   - Variable components are mainly used to provide variety (diverse specifications) in line with buyer needs, and are therefore established as variable components in products. (These components are used in specialized ways according to the product type and other product specifications).

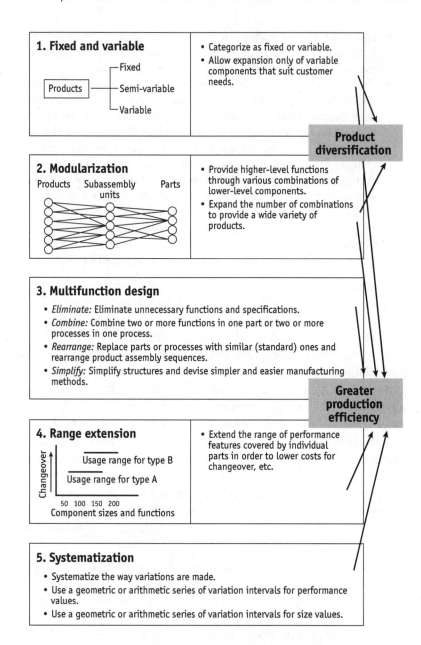

Figure 4-29. Techniques for Reevaluating from the Product Design Stage

- Meanwhile, try to use as many common parts and materials as possible for fixed components in order to reduce the variety of parts and materials. In addition, rationalize and simplify manufacturing processes as much as possible to facilitate manufacturing and to introduce more automation when it is appropriate.

2. *Modularization.* Examples of modularization include parts and subassembly units used in automobiles and Lego-style toys. There are two things you can do.
   - Simplify parts and units, and then use combinations of simplified parts to provide a wide variety of product specifications.
   - For product configurations, create advanced functions using relatively simple lower-level components.

3. *Multifunction design.* Review products and production methods in terms of functions and structures and look for ways to achieve similar functions and structures using fewer parts or fewer (or simpler) manufacturing processes. Pay special attention to the ECRS principles of improvement.
   - Eliminate unnecessary functions and specifications.
   - Combine two or more functions in one part or two or more processes in one process.
   - Replace parts or processes with similar (standard) ones and rearrange product assembly sequences.
   - Simplify structures and devise simpler and easier manufacturing methods.

4. *Range extension.* Extend the range of performance features covered by individual parts in order to lower costs for changeover, equipment, dies, jigs, tools, etc.

5. *Systematization.* Systematize the variations in dimensions or structures by implementing a set of rules for relevant specifications (for equipment, dies, jigs, tools, etc.) and manufacturing conditions so that management becomes simpler (more systematic). For example, use a geometric or arithmetic series of variation intervals for sizes or performance values, such as the metric and imperial systems that are used for the sizes of nuts, bolts, etc.

When a company applies these five techniques to reduce the variety of purchased parts and materials, as well as the variety and complexity of product specifications and manufacturing processes, the purchasing of parts and materials becomes much more conducive to synchronization and equalization and opens the way for the company to improve its manufacturing system to take advantage of indirect and comprehensive cost reductions. With the ESP approach, analysis of existing products is thought of as a starting point for making improvements in parts and materials that will lead to major long-term benefits while you make impressive innovations in manufacturing and in the design of new products.

The types of production system innovations we've been describing are essential for success in creating an ESP Production System. And they have also been very beneficial to companies that have already introduced an ERP (Enterprise Resource Planning) or SCM (Supply Chain Management) system. As was explained in Chapter 2, many companies are using ERP or SCM to operate a ready-made production management system, and as a result they have had a lot of problems trying to fit their particular operations into the ready-made mold. This is largely because these approaches focus mainly on information systems and hardly touch upon factory-floor problems. No matter how good your company's information system is you cannot hope to make its production operations more efficient and free of confusion unless your production can operate on schedule and with accurate input provided by the factory. That is why manufacturing process innovations play such a key role in making the ERP and/or SCM approach a success.

In short, manufacturing innovations are the results of steady, persistent activities that you must implement to the utmost if you want to build up strong support for your company's manufacturing business.

KEY POINTS FOR BUILDING AN ESP INFORMATION SYSTEM

Figure 4-30 outlines the basic approach for building an ESP information system. As for the first point (that the ESP information system will coexist with the company's current computer system), these days almost every manufacturing company has introduced a computer system for production management based on MRP. Given the immense amounts of data that each of these systems uses, it would be very expensive to replace them with a different computer system.

Consequently, the ESP Production System will coexist with the current computer system and will basically augment that system with some new planning functions. To accomplish this, the planning standards must accurately reflect the capabilities (results) of the company's manufacturing processes. To ensure this, you must establish a results-data collection system, and it must include functions for periodically revising the planning standards to keep them up to date with daily production results. This results-data collection system does not have to be a complicated system if you have already established ESP. But this information system will make previously obscure production-related problems more visible, will help

---

Basic approach to build an ESP information system

1. Assume that the ESP information system will coexist with the company's current computer system.
2. Establish an accurate results-data collection system.
3. Develop a program that is geared toward the company's own characteristics.
4. Assume that it will include personal computers and will enable flexible responsiveness.

Figure 4-30. Basic Approach to Building an ESP Information System

to identify the improvement themes that promise the greatest benefits for the company, and will help improvement activities in general remain oriented in the most appropriate direction. At the same time, it is important to monitor and control the ways in which you implement improvement activities. The following describes an ESP information system, which makes use of the five techniques of synchronized and equalized scheduling that were explained earlier.

A Support System for Production Planning Proposals

The ESP information system is basically a support system for production planning proposals. As such it includes functions supporting the proposal of the main schedule, semidetailed schedule, and detailed schedule. These functions are selected according to the range of time to be covered by the proposed production schedule. So the first step is for ESP's information system autoproposal functions to apply actual, confirmed order information to preset production patterns to create a standard-based production schedule. The load is adjusted in the main and semidetailed schedules in accordance with the characteristics of the company's production system.

The next step is to conduct various reviews of the standard-based production schedule, including an overproduction prevention review, zero missing items review, and load feed-forward review, which we discussed earlier in this chapter. These three reviews help optimize the arrangement of production patterns as part of the auto-adjustment process for production schedules. Figure 4-31 shows a flowchart of the autoproposal processes. The overproduction prevention function, zero missing item function, and load feed-forward function that are among these auto-adjustment functions can be used to check other specific adjustment functions after users have made their own adjustments.

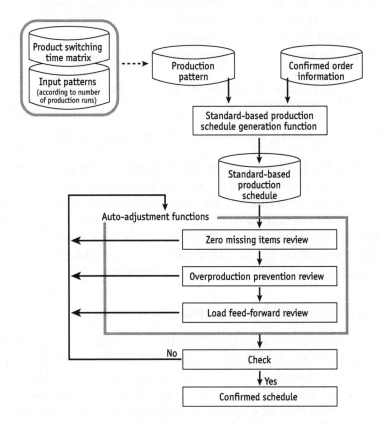

Figure 4-31. Flowchart of Autoproposal Processes

Figure 4-32 shows an example of the logic behind these processes. The word 'blank' is used in this figure to indicate instances of excess capacity. During the zero missing items review, you make insertions wherever missing items are found and then you perform an overproduction prevention review to check for the blanks where the insertions will go. Accordingly, these two reviews are interrelated.

As was mentioned earlier, each company's ESP Production System is built somewhat differently according to the company's own characteristics. For example, each company uses different production patterns and even its current computer system will uniquely reflect the company's characteristics to some extent. Although the ESP information system uses personal computers and its own set of

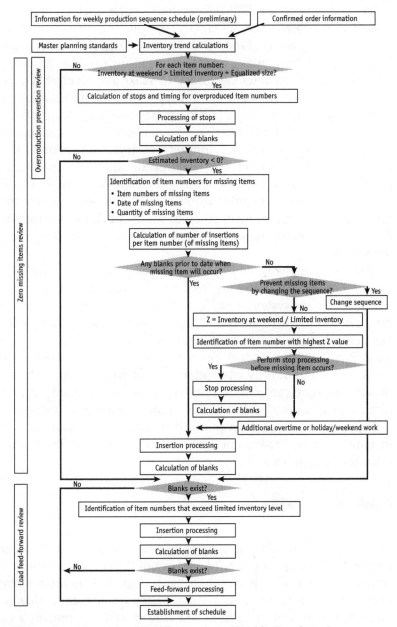

Figure 4-32. Example of Simulation Steps (Weekly Simulation)

software, there are some companies whose current computer systems include software that can be adapted for use in support of the company's ESP information system.

The contents of the ESP information system include more than information related to the company's production management system. Accordingly, the standard software on each company's computer system often cannot be readily used for the ESP Production System. Thus it becomes necessary to make improvements in the computer's software so that it will support an ESP information system and suit the company's particular needs and characteristics.

Typically, the ESP information system operates on a system comprised of personal computers running a customized version of Excel under Windows as the primary application and GUI (graphic user interface). Excel was selected because it enables users to draft high-quality production schedules without having to perform any complicated programming. Also, much of the master data is maintained on a host server to avoid having to replicate the master data on numerous PCs.

Under this network configuration, each PC can download data needed for drafting schedules from the network's host server in the ESP information system. Once a schedule has been completed, you can then transfer it to the host computer so that the production staff can access the production specifications as online information. With an ESP information system, users can select menu buttons on the Excel screen and directly edit various values displayed on the screen. In addition, they can plug in revised values as confirmed values for recalculated autoproposals and use various adjustment functions.

This concludes our description of the steps and techniques used to build an ESP Production System. Let's look at a few common mistakes companies make.

COMMON MISTAKES

If there is one common mistake our experience has shown, it is that employees at most of these companies tend to neglect, or do not feel compelled to make, any careful analysis of the current conditions at their companies. This is largely because they feel that, we already know what's going on here, we don't need to analyze it. However, no matter how well one thinks one understands a company's current conditions, there are always many details that remain undiscovered. Employees are always surprised at how much they find out by doing these analyses. When attempting to build up a new system, it is especially important to study and critically review the way things are.

Another common mistake is when managers try to improve their production management system without including manufacturing process improvements. It should be quite clear to anyone who has read this book thus far that no ESP Production System can succeed unless it includes manufacturing process improvements. The authors therefore ask that readers follow *all* of the steps described in this chapter so that they can make steady progress toward success.

5

ESP Result Indicators

In many cases, when we've talked with employees from companies that have implemented various improvement activities such as building new systems or functions, they have a lot to say about the activities themselves, but very little to say about the results. Generally, the problem at such companies is that the goal of their activities has been to build certain new systems or functions, rather than to achieve certain results. In such cases, the people involved in making improvements haven't set quantitative goals, or haven't been systematic in deploying result indicators (that is, they have not established clear improvement stories for their improvement activities). Or, they have planned top-heavy improvement projects that sound good in theory, but do not follow through with improvements at the practical level of actual work processes.

DEVELOPING ESP RESULT INDICATORS

When evaluating how well you have implemented ESP we look for specific results. (See the case studies described in Chapters 6 through 9.) Part of building an ESP Production System is making quantitative analyses of goals and results. You achieve results through the Production Division's concerted efforts to set targets, deploy targets as improvement projects, and evaluate the completed improvements.

ESP is a program that concurrently involves manufacturing process improvements and the establishment of a new production management system that has three principal objectives: 1) boost productivity, 2) achieve all delivery deadlines (zero missing items), and 3) reduce (minimize) inventory. These are the three areas where we mainly apply result indicators (top target indicators).

At the level of the company's business results, the result indicator for productivity is reflected in the reduction of product costs through higher productivity, while the result indicator for reduced inventory is reflected in a reduction of current assets and improved cash flow. In addition, the achievement rate for all delivery deadlines (zero missing items) by meeting the buyer's required lead times (or shortening delivery lead times when necessary) works to raise customer satisfaction levels.

The quantitative values of the target improvement levels for the three principal goals described above are set only after examining and coordinating with the company's business policies and goals, the product delivery requirements of buyers, and with the benchmark comparison with competing companies. The key to successful improvement activities lies in getting everyone in the company to understand fully the need for reaching the target levels, and the benefits gained by doing so. This way they can share a solid commitment to work together for success.

For example, the case study in Chapter 7 concerns the Okegawa plant of a Japanese company called Izumi Industries, where an ESP Production System was built to reduce inventory loss as the specific goal for cost-cutting activities. The essential requirement in developing their overall improvement activities was to reduce inventory and raise productivity.

Although we must take into consideration factors such as the type of industry and the actual requirements when setting target values, using our experience-based guidelines we set goals of 1) a 1.5-fold increase in productivity, 2) 100 percent on-time delivery (zero missing items), and 3) inventory reduction of at least 10 percent. The indicators for these three principal goals are:

- Higher productivity.
- Labor productivity, measured as output per labor hour or unit of labor cost.
- Productivity of planning and management tasks, measured as time required for production planning tasks, etc.

- Equipment productivity, measured as process output, equipment's total efficiency, etc.
- Achieve all delivery deadlines (zero missing items).
- Achievement of schedule for specific item numbers, measured as full production of item numbers as scheduled.
- Achievement of production output for specific item numbers, measured as full production of amounts as scheduled.
- Achievement rate for delivery lead times per buyer delivery lead time (period from reception of order to delivery to buyer).
- Process-specific achievement of production as scheduled, percentage of production schedule achieved.
- Achievement of on-time delivery of purchased parts (etc.).
- Inventory reduction (minimization).
- Inventory reduction rate per product, in-process inventory item, part, etc.

The chart in Figure 5-1 illustrates the system of ESP result indicators and shows the relationships between result indicators and activities. The result indicators for goals and targets are on the left side of the chart and those for methods and policies are on the right side. Items in parentheses refer to the considerations made when building an ESP Production System.

As we mentioned earlier when describing manufacturing process improvements, boosting process capacity at bottleneck processes (or equipment units), totally eliminating sudden breakdowns or other abnormalities that interfere with the production flow through processes, and minimizing changeover and adjustment times to enable smaller lots, are all fundamental for the ESP Production System.

The goals of raising process capacity at bottleneck processes and minimizing changeover and adjustment times are deployed when establishing a synchronized and equalized production system. In other words, you set each goal, not simply in terms of boosting productivity, but rather in terms of eliminating bottleneck processes that would otherwise occur when the number of changeovers increases as a result of production in smaller lots (see Figure 5-2). Once you have clearly identified the target bottleneck processes and/or equipment units and set improvement targets for each of them, it is important to start making your improvements on the highest priorities first. After starting these improvement activities, you may still have to revise the prioritization list from time to time, since factors such as the product configuration, the production volume, and the product models currently in production

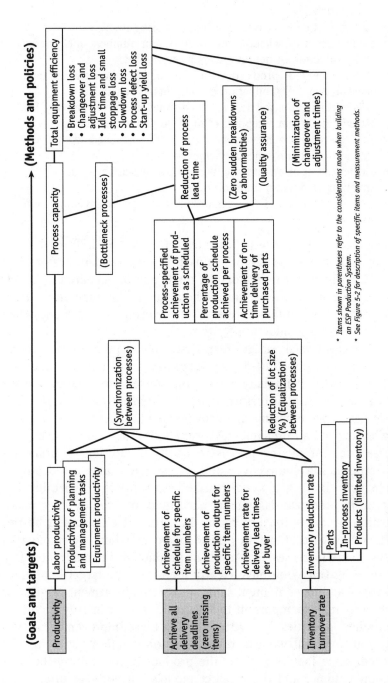

Figure 5-1. System of ESP Result Indicators

* Items shown in parentheses refer to the considerations made when building an ESP Production System.
* See Figure 5-2 for description of specific items and measurement methods.

may cause the required level of improvement to change for certain bottleneck processes or equipment units.

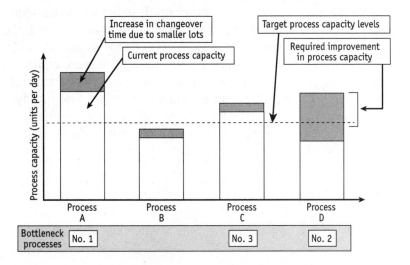

Figure 5-2. Deployment of Improvement Goals (by Priority) Among Bottleneck Processes

The result indicators in Figure 5-1 are mainly the responsibility of the Production Management Division, not the factory-floor staff, and it is important that the production managers take the perspective of optimizing the manufacturing system as a whole when deploying targets and proposing improvement themes. Production managers are charged with establishing management (result) indicators, such as the achievement rate for delivery lead times and the inventory turnover rate, as well as for creating a systematic deployment of goals and building a system for collecting and managing the results.

The Production Management Division is also the main player when it comes to setting up improvement activities and systematically checking their results by analyzing how thoroughly they have achieved the improvement plan. They must also promote thorough evaluation of improvement activities in general. To do this, production managers should work daily to stay abreast of activities in the factory, using graphs and other tools to clarify deadline-related goals and to get a firm handle on whatever problems may arise.

Working with ESP Result Indicators

The result indicators listed in table 5-1 are all basic ESP indicators. You should study and establish other result indicators when appropriate for the current improvement theme goals (such as reducing overtime hours, sudden breakdowns, and/or process abnormalities) as well as target items, product models, parts, processes, equipment units, and so on. Also, additional formulas, units, and management cycles used with measurement methods can be set after studying these other types of result indicators. See Table 5-1 for some examples of measurement methods used with result indicators.

Table 5-1. ESP Result Indicators (Examples)

Result indicator		Measurement method (example)			
Category	Item	Formula	Unit	Mgmt. cycle	Remarks
Productivity	Labor productivity	Value of production / Total labor hours of production-related employees	$ per hour	Annual or semiannual	
	Direct labor productivity	Value of production / Total labor hours of direct production-related employees	$ per hour	Annual or semiannual	
	Productivity of planning and management tasks	Value of production / Total labor hours of planning and management staff	$ per hour	Annual or semiannual	
	Process capacity	Units that can be produced per day (or units per month)	Units per day	Annual or semiannual	Bottleneck processes
	Overall efficiency of equipment	Logical cycle time x Number of nondefective units/ Load time or (time-based operation rate) x (Performance-based operation rate) x (Nondefective rate)	%	Monthly	For each target equipment unit
Deadlines	Achievement rate for delivery deadlines (delivery achievement rate)	No. of on-time deliveries / Total number of deliveries	%	Monthly	
	Achievement of schedule for specific item numbers	Actual amount of production item numbers / Amount of item numbers in production specifications	%	Daily or weekly	
	Achievement of production output for specific item numbers	Actual production volume per item number / Scheduled production volume per item number	%	Monthly	
	Achievement rate for delivery lead times per buyer	Standard delivery lead time (before improvement–after improvement)	Daily	Annual or semiannual	
	Process-specific achievement of production as scheduled	Actual amount of production item numbers / Amount of item numbers in production specifications	%	Daily or weekly	For each target process
	Percentage of production schedule achieved	Actual production volume per item number / Scheduled production volume per item number	%	Monthly	
	Achievement of on-time delivery of purchased parts	No. of purchased parts delivered on time / No. of purchased parts ordered	%	Monthly	For each supplier
	Process lead time	Scheduled standard lead time or actual lead time	Daily	Annual or semiannual	
	Lot reduction rate	No. of changeovers after improvement / No. of changeovers before improvement	%	Monthly	
Inventory	Inventory turnover rate	Value of inventory (products, in-process inventory items, and materials) / Value of shipped products	Monthly	Annual or semiannual	For each product, in-process inventory item, or material
	Inventory reduction rate	(Before improvement–after improvement) inventory turnover rate / Inventory turnover rate before improvement	%	Annual or semiannual	

Case Study 1:
Miyagi Shimadaya Co., Ltd.

Our new production management system boosted
equipment efficiency rate into the 90 percent range!

How ESP brought high efficiency to a food manufacturing line
Building a production management system and line improvements
for better responsiveness to daily orders—

Miyagi Shimadaya Co., Ltd.

In Japan, noodles are a traditional staple food much like rice and
bread and, in fact, Japan's noodle sales are estimated at over 1 tril-
lion yen (US$8 billion) per year. Noodles are sold in Japan in four
basic categories: udon, soba, Chinese noodles, and pasta. These
days, Japanese consumers show a marked preference for high-
quality noodles, especially fresh noodles. They also like noodle
products that are easy to prepare.

Even in a market inhabited by such discerning consumers,
Miyagi Shimadaya Co., Ltd. has done well by developing
advanced technology in noodle manufacturing. In fact, they have
been a pioneer in Japan's boiled noodle industry. In recent years,
the worldwide "sushi boom" has spurred the popularity of Japanese foods in general in many countries, and noodles have joined
sushi and other Japanese foods as a popular choice among health-
minded consumers. Currently, Miyagi Shimadaya exports noodles
to various overseas markets, including Hong Kong, Britain,
Australia, and the United States, and growth has been particularly
strong in exports to Britain. These export products are mainly
dry noodles that can be stored at room temperature and frozen

noodles. Most are either udon-style noodles (including yakiudon for frying) or Chinese-style noodles (including yakisoba and ramen types).

COMPANY CHARACTERISTICS AND DISTRIBUTION SYSTEM

Miyagi Shimadaya's best-selling products are dry noodles (for boiling) that are packaged to remain fresh at room temperature for 100 days. These products appeal especially to consumers who like noodle dishes that are easy to fix, quick, and delicious. For a long time now, Miyagi Shimadaya has been able to use the long 100-day shelf life of these products to its advantage by running its production system systematically and efficiently. However, the company's delivery system for these products is the same as for refrigerated noodles (with freshness periods of 4 to 15 days) that are produced by other companies in the Shimadaya group. As a result, the dry noodles are distributed to retailers via the same distribution operations that are used for perishable, refrigerated noodles.

Overview of Manufacturing Processes

Miyagi Shimadaya's manufacturing processes are generally divided into the following five types (see Figure 6-1). Miyagi Shimadaya has introduced an HACCP system to help prevent hazards at every stage from receiving raw materials to consumption of final products.

Primary packaging processes

1. *Noodle-making processes.* After the raw materials (flour, etc.) are mixed and shaped into sheets, the sheets go through several roller mechanisms to make them longer and thinner. A rolling cutter then slices them into ribbons (noodles) and also cuts them to the required length before they are passed to the next process. The manufacturing steps so far take from 40 to 50 minutes. Although the manufacturing steps themselves are automated, a great deal of human labor is still required for tasks such as switching the materials and setting up and cleaning the equipment.
2. *Boiling processes.* Each serving of noodles, which has been cut to the standard length at the end of the noodle-making processes, is inserted one at a time into a case, which is then placed into a hot

1. Noodle-making processes
2. Boiling processes
3. Filling processes
4. Sterilization and cooling processes
5. Secondary packaging processes

Processes at 1 to 4 are called the "primary packaging processes"
Processes at 5 are called the "secondary packaging processes"

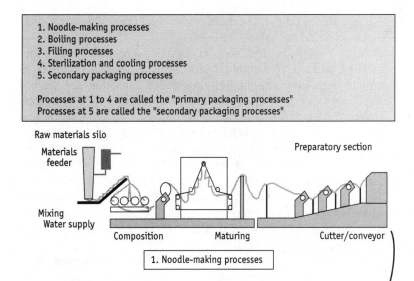

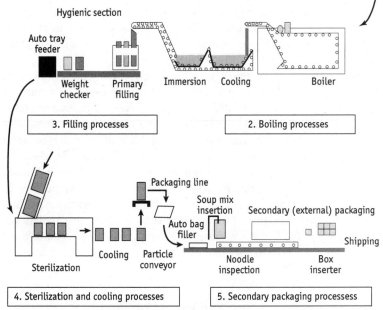

Figure 6-1. Manufacturing Processes

water vat for boiling. After boiling, the noodles are force-cooled with water, then conveyed to the next process. The boiling processes take about 20 minutes up to this point. Like the noodle-making processes, these boiling processes are automated, but still require a lot of human labor for changing materials, setting up, cleaning, and so on. In other words, the time loss incurred by these manual tasks is the main reason why the equipment's over-all operation rate is low.

3. *Filling processes.* The noodles, which have been boiled and cooled in units of individual servings, are inserted by an auto-matic bag filler into single serving-size bags that are marked with certain symbols. Then, each bag is hermetically sealed. Next, the bagged noodles are sent through a weight checker and a metal detector. The weight checker discards any underweight packages and the metal detector makes sure the bags do not contain any metal particles. After that, the auto tray feeder arranges the noodle-filled bags onto 12-unit trays (3-4 rows each) and then the trays are conveyed to the sterilization and cooling processes. These processes are also automated, but human operators are needed to monitor the equipment and visu-ally inspect the products.

4. *Sterilization and cooling processes.* The trays sent from the filling processes are stacked ten high, and then the stack of trays is sent through the sterilizer (steamer). Next, the trays are cooled to room temperature by a blower on the conveyor, then sent to the sec-ondary packaging processes. The sterilization and cooling processes are fully automated (unmanned).

5. *Secondary packaging processes.* The single-serving bags of noodles on the 12-serving trays are automatically removed in groups of four bags and are placed on a conveyor. Next, a human operator picks up the bags from the conveyor and places them individually onto the packaging conveyor. On the packaging conveyor, each bag of noodles receives a small envelope of broth mix, then the bag and envelope are wrapped in plastic. After that, a human operator inserts 10 noodle packages into each box. When three of these 10-package boxes are filled, they are packaged again as final products. These final products are then loaded onto pallets for shipment. Thus, the secondary packaging processes require a lot of human labor, in fact, about twice as much as do the primary packaging processes.

OVERVIEW OF CONDITIONS AND PROBLEMS PRIOR TO INTRODUCING ESP

This overview covers the stages of reception, scheduling, produc-tion, and shipping prior to introducing the ESP Production System. All customer orders are received by the Shimadaya Group's head office, which distributes the orders to the relevant group companies, including Miyagi Shimadaya. Miyagi Shimadaya then proceeds

1. Lead time required for primary packaging processes and other related issues.

2. A little more than three hours are required for the noodles to complete the primary packaging processes. Idle time occures at the secondary packaging processes whenever the secondary packaging processes are ready to accept more trays of noodles but the trays of noodles have either not been inserted into the primary packaging processes or have not yet completed those processes.

3. In addition to the problem of a 3-hour lead time for the primary packaging processes, other problems arise due to the fact that the primary and secondary packaging processes are not directly linked as continuous processes.

   Consequently, there is often an inventory of semifinished products (i.e., trays of noodles that have completed the primary packaging processes) that is retained between the primary and secondary packaging processes.

   The existence of an inventory of trays of noodles between the primary and secondary packaging processes creates the need for human labor for inventory management, such as loading and unloading.

with production based on the orders it has received. Figure 6-2 shows a flowchart of the main stages of production at Miyagi Shimadaya, from reception of orders to shipment of products, prior to the introduction of an ESP Production System. Each week, Miyagi Shimadaya receives the orders for the following week's production, at which point the production managers draft a production schedule based on the order. Next, they draft a shipping schedule, followed by a secondary packaging schedule and primary packaging schedule in that order.

Since this company's products are household food items (noodles), the popularity of specific types (and amounts) of products varies greatly according to factors such as the weather and the season. Consequently, it is not easy to estimate demand in advance, and changes in orders received the week prior to production are frequently changed. Consequently, Miyagi Shimadaya was used to having daily changes made in its secondary packaging schedule and primary packaging schedule, and it came to regard all of its weekly schedules as approximate rather than exact planning.

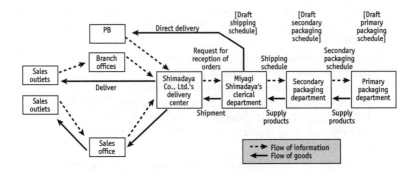

Figure 6-2. Flow from Order Reception to Shipment

Problems with the Production Packaging Schedules

Since the weekly production schedules (primary and secondary packaging schedules) varied from day to day, the company was unable to order materials based on a reliable weekly production schedule, and so there was a lot of guesswork involved when ordering materials. As a result, there were days when shortages of materials occurred, which meant they had to stop the production line and/or rearrange the sequence of processes.

Under these conditions, the production schedule did not become confirmed until the day of production, after the last-minute changes had been made in the schedule. Thus, production planning in effect was being done on a daily basis, which meant that the production volume varied from day to day. On one day the production line might have to operate overtime to meet an extra large order, and the next day it might be possible to fill the day's orders by noon. The day-to-day production load was anything but even.

Having to accommodate last-minute order changes (spot orders) also resulted in frequent rearrangement of production processes and delays in getting materials needed for products, which meant that various production equipment units would sit idle. Miyagi Shimadaya decided it was time to get rid of this inefficient production system and replace it with an ESP Production System in hopes of making big improvements in efficiency. Figure 6-3 summarizes the problems faced by Miyagi Shimadaya.

This company manufactures products with two brands: the Shimadaya brand ("NB" or native brand items) and a private brand (PB items) on an OEM basis for certain customers. In addition to the various problems listed in Figure 6-3, Miyagi Shimadaya had other problems arising from the fact that there were separate ordering systems for NB and PB items. Since NB items were ordered via the

	Problem	Description
Order reception problems	1. Lots of loss due to schedule changes	Problems due to daily variation in orders: • Production has to accommodate last-minute additions and cancellations from customers, which means having to change products on the line more often and therefore more loss due to frequency setups, etc. • Also, this resulted in other types of loss, such as equipment stoppage loss, changeover loss, and management loss.
	2. Leveled production not possible	Problems due to producing and shipping orders received yesterday: • When today's production is based on orders received just yesterday, the amount of orders can vary widely from day to day and thus the production load also varies proportionately, so that it was not unusual if there happened to be an overtime load one day and less than a full day's load the next day. • Under these conditions, it is not possible to level production and therefore equipment and labor are not always used efficiently.
Equipment-related problems	3. Primary packaging processes are inflexible	The period from feeding materials to the production line to reaching the secondary packaging processes is a little more than three hours: • Therefore, the semifinished products are not always available for the secondary packaging processes when they are needed. In such cases, the secondary packaging processes are frequently changed over to whatever semifinished products are available, which frequently incurs stoppage-related loss. • Due to the daily variation in orders, the inventory of semifinished goods may contain semifinished products that are not needed in addition to the ones that are needed. Unbalanced inventory: As the inventory of semifinished products grows, shortages of product trays and other problems can occur due to poor circulation of resources.
	4. Secondary packaging processes are inflexible	Problems with the equipment on the packaging line: • There are three secondary packaging lines, and only certain types of products can be packaged on each line. Consequently, the operation time of each line varies greatly depending on the types and amounts of products ordered each day. • Even when the daily total production volume for ordered products per day is less than the factory's capacity, one line may be underutilized while another line is being operated overtime to handle its heavy production load. • Therefore, equipment and labor are sometimes used very inefficiently.

Figure 6-3. Problems Before Introducing ESP

Shimadaya Group's branch offices and sales office, the company could expect to receive advance order information that were reliable to a certain extent, and could therefore set up some degree of advanced production scheduling. However, the order information for PB items came later, leaving no room for such advanced planning. The main reason for this difference is that all orders for PB items were routed through the Shimadaya Group's head office, often via a long, complicated path of communications. Consequently, many of the last-minute changes that were made in the weekly production schedule were for PB items.

Therefore, we began by analyzing NB items and PB items with a view toward somehow equalizing the flow of orders. Figure 6-4 shows an illustration that resulted from this analysis. As you can see, about 60 percent (by volume) of the ordered items could be produced on a reliable weekly schedule, and all of these items are NB items. The remaining 40 percent were spot orders that were subject to daily changes, and of these 20 percent were NB items and 20 percent, PB items. Therefore, before ESP was introduced, the company had to accommodate daily changes for about 40 percent of its products. However, sometimes daily changes had to be made even for the other 60 percent in order to ensure on-time delivery.

Given this situation, how could the company turn out products in a way that was flexible toward daily order changes and without having any missing items? Furthermore, how could the company operate its production lines at anything close to maximum efficiency? These were tough problems indeed.

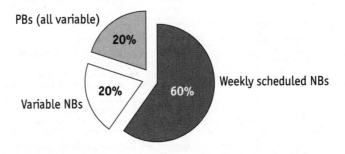

Figure 6-4. Product Breakdown: Weekly Scheduled Products and Variable Products

MIYAGI SHIMADAYA INTRODUCES THE ESP PRODUCTION SYSTEM

Before introducing ESP, each company should reach a consensus on a vision: what kind of production system is best, and how they should pursue it. This consensus should be focused on improving the production system so that it can accomplish all of the following basic needs.

- Preserve what is best about its products.
- Deliver products to customers on time.
- Give customers just what they want.
- Deliver only the amounts needed by customers.
- Ensure the quality of its products.

In other words, the company must find a way to deliver products of assured quality to its customers by the agreed-upon delivery deadline, which is something that the company has attempted to do in the past by waiting until the previous evening to finalize the next day's production schedule. This consensus concerns only the what, not the how, of the desired production system, ESP. The achievement of this vision requires teamwork on the part of everyone, from line workers to shipping clerks to top managers at the head office.

The four problems described in Figure 6-3 became evident after studying how the production load scheduled for the secondary packaging processes varied widely from day to day, due to the last-minute finalization of the shipping schedule to accommodate a set of spot orders that had been subject to change right up until the evening of the previous day. Again and again, this wide variation in the production load for secondary packaging processes had the following adverse effects on the primary packaging processes.

Examples of adverse effects on primary packaging processes

1. Sometimes, semifinished products that had come off the primary packaging processes the day before were no longer needed because the orders for them were canceled on the next day.

2. Conversely, additional orders made at the last minute might mean that certain products not originally scheduled for the secondary packaging processes must be added as products to be manufactured via the primary packaging processes on the same day.

To deal with these types of last-minute order changes, Miyagi Shimadaya would:

- Manufacture products to be shipped that day as specified.
- Find ways to revise the shipping schedule and production schedule at the same time.

Miyagi Shimadaya had been doing this for many years, so they were used to having to deal with large amounts of variation in production load and, consequently, sharp reductions in production efficiency. Unfortunately, this arrangement did not enable them to take advantage of their products' relatively long shelf life.

Their next step was to find some way to have an efficient production management system that made the most of their products' characteristics while dealing with orders that, as was shown in Figure 6-4, included 60 percent whose production could be scheduled in advance and 40 percent whose production remained variable until the last minute. We will now look in more detail at the four stages in establishing an ESP Production System.

Stage 1: Easing Constraints on Shipping

The first order of business was to find ways to alleviate the constraints that existed for the company's shipping operations. The fact that Miyagi Shimadaya's food products have a relatively long shelf life makes improvement of the distribution system especially important, but whatever improvement is made must be a top-down improvement involving the company's upper management.

After conducting analyses, it was decided that the best approach would be to establish a product inventory of NB items in order to provide the needed flexibility in the scheduling of shipments. After holding discussions with managers at the Shimadaya head office, the Miyagi Shimadaya launched an improvement project. Until then, shipping operations were constrained by a rule based on the belief that products with older dates of manufacture will not sell easily. In other words, since products are sent into and out of inventory on an FIFO (first in, first out) basis, the products with the most recent dates of manufacture are not shipped as long as previously manufactured products remain in the product inventory, and therefore the shipped products tend to be older ones that are harder to sell.

To avoid this problem, the company had treated their dry noodle products (shelf life: 100 days) the same as the raw noodle products (shelf life: 4 to 15 days) by establishing a rule that "all products must be shipped on their date of manufacture." Accordingly, they had to schedule production based on this rule. However, the Shimadaya head office staff conducted an opinion survey among consumers and discovered that the vast majority of the survey respondents who bought Shimadaya dry noodles were well aware of their long shelf life. The Miyagi Shimadaya therefore decided it was safe to conclude that dry noodle products could be shipped from product inventory after the date of manufacture without having to worry about a significant decline in sales.

As a result, the staff at Miyagi Shimadaya and at the Shimadaya head office decided it was feasible after all to set up a product inventory of NB items at the factory. Thus, the shipping operations were changed as follows.

Constraints on shipments of NB items	
● *Before improvement*	Products are shipped on date of manufacture (date of manufacture = date of secondary packaging process) or on the day after the date of manufacture.
○ *After improvement*	Products are shipped from product inventory on FIFO basis.

In order to implement the new method of shipping products from product inventory on an FIFO basis in a way that fully complies with the company's product quality assurance policy, they set standards for production periods and limited inventory (maximum inventory level for each product item). Next, at stage 2, the main task would be to build a new production planning system that would be fully adaptable to daily variation in orders by using the product inventory as a buffer to enable efficient, scheduled production.

Stage 2: Pursuing Leveled Production and Setting Limited Inventory Levels

The first question to be answered at Stage 2 in establishing ESP was "Is leveled production really possible?" Stage 2 was also the stage for setting limited inventory levels. The company looked into the potential for leveled production by examining the amounts shipped per item each day and the amount of variation from weekly schedules during the previous six months. Figure 6-5 illustrates the results of analyses in which these shipment amounts and variation were converted to load time values.

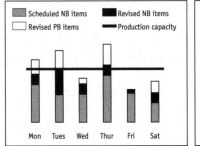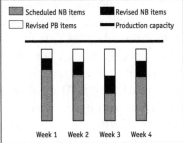

Figure 6-5. Daily and Weekly Shipment Trends (Converted to Load Time Values)

The results of these analyses showed that, although there was considerable variation from day to day, the level was fairly even from week to week. (In Figure 6-5, the six months of data was collected on the day following each production date and the monthly data was divided by four to yield weekly data.) The results also indicated that, in weekly units, there was a sizeable surplus of production capacity.

As mentioned above, the results clearly show a large amount of variation from day to day. Indeed, the data collected on each day following the manufacture showed that on some days the production load exceeded the production capacity. Nevertheless, when the results were leveled out on a weekly basis, production capacity always exceeded the production load. The results of the analyses of past orders provided clear evidence that leveled production would be possible if NB products were to be shipped from product inventory on an FIFO basis, as was proposed at Stage 1, as a way to alleviate the constraints that existed for the company's shipping operations.

Next, they used the above analysis results to set limited inventory levels (appropriate levels for product inventory) for NB items.

These analysis results were also used to set an orientation for solving the various problems that were listed in Figure 6-3. The orientation they took in responding to these problems is described in Figure 6-6.

	Problem	Orientation of improvement
Order reception problems	1. Lots of loss due to schedule changes 2. Leveled production not possible	Order changes will not be immediately reflected in the production schedule. 1. The shipping schedule will vary depending on changes made to orders in the weekly schedule. However, changes in the shipping schedule will not require immediate changes in the production schedule (for secondary packaging processes). (The shipping schedule and production schedule will not be linked unidimensionally.) 2. The production schedule will be drafted based partly on the equipment's production capacity. (The standard will specify a production capacity utilization of 100 percent in order to maximize production efficiency.) 3. Time segments to be used for variable elements (estimated variation in orders) will be worked into each production schedule. 4. Products with high turnover rates will be used to adjust the load time when drafting production schedules. (A product inventory of products with high turnover rates will be maintained.)
Equipment-related problems	3. Primary packaging processes are inflexible	Production scheduling will include specified amounts of semi-finished products. 1. In order to prevent missing items at secondary packaging processes, a limited inventory (with specified upper and lower limits) of semifinished products from primary packaging processes will be determined and maintained. 2. The production schedule for primary packaging processes will not be immediately changed according to last-minute changes in orders. 3. Currently maintained levels of semifinished products will be considered when planning the production schedule for primary packaging processes. 4. However, the production schedule for primary packaging processes will mainly be based on the production schedule for secondary packaging processes.
	4. Secondary packaging processes are inflexible	Development of flexible packaging lines. 1. Switch from product-specific specialized secondary packaging processing lines to secondary packaging processes that share the same equipment. (All secondary packaging processes should be able to handle all product types.)

Figure 6-6. Orientation of Improvements in Response to Problems Existing Prior to Introducing ESP

Stage 3: Improving the Production Planning Methods and Establishing the Production Scheduling Steps

The first step in drafting a production schedule for manufacturing processes is to set a schedule for the secondary packaging processes. After that comes the production schedule for the primary packaging processes. Accordingly, improvements in production planning should start with improvement in planning production for the secondary packaging processes.

The approach to making improvements is one of devising a production planning method (for secondary packaging processes) that takes into consideration the points listed in Figures 6-3 and 6-6. This means that the primary focus will be on devising improvements to eliminate the major forms of loss that have plagued the company's manufacturing operations: changeover loss and loss due to variation in operation time (the latter includes loss due to overtime, underutilization of production capacity, and idle time). We will now discuss five improvements.

Improvement 1: Use limited inventory to separate order changes from production schedule for secondary packaging processes. The company's production schedule before improvement was drafted in three steps—shipping schedule, then secondary packaging production schedule, then primary packaging production schedule—nominally based on a week's worth of orders. However, the production specifications for the manufacturing processes were actually based on the orders for product items and quantities that were confirmed just the day before, and the day's shipping schedule, secondary packaging production schedule, and primary packaging production schedule had to be changed accordingly. This arrangement resulted in loss due to product switching and inefficiency due to variation in production load, as was described in Figure 6-3.

The improvement-making approach was described in Figure 6-6, which states that order changes will not be immediately reflected in the production schedule and changes in the shipping schedule will not require immediate changes in the production schedule (for secondary packaging processes). To make these improvements, products with high turnover rates will be used to adjust the load time when drafting production schedules (also mentioned in Figure 6-6). Therefore, order changes will be separated from the production schedule for secondary packaging processes.

Consequently, it was decided that a product inventory, consisting of products with high turnover rates, would be established as limited inventory of NB items, based on the results from analyses of orders received over six months.

Improvement 2: Setting a product input sequence (for production) that minimizes changeover loss. Previously, the product items placed first in the production sequence by the production specifications were those product items that had the greatest quantities in the previous day's confirmed orders. This resulted in quite a number of product changeovers and switching of supplied materials, which meant a great deal of time-related loss. Whenever products were switched at the secondary packaging processes, it required setup of new semifinished products and materials for the new product, as well as various equipment-related changeover operations. Figure 6-7 lists the various factors and problems that led to changeover loss in secondary packaging processes before the company made improvements.

A study of these changeovers that were being made prior to the improvements revealed that the same semifinished products and materials were being set up for the line several times in the same day. Looking into the factors behind this, it was found that different product items that use the same semifinished products and materials were being produced at various times of the day, but that fact was never taken into account when planning production, which was based strictly on the product items' specified date of manufacture and quantities to be shipped. Figure 6-7 also lists the production input sequence that was used prior to improvement.

In other words, production specifications were determined for the secondary packaging processes based on the criteria described in Figure 6-7 and without considering the impact on frequency of changeovers. Consequently, it turned out that semifinished products and materials had to be switched whenever product types were switched. Thus, the same semifinished products and materials were being delivered several times in the same day, resulting in frequent changeovers (definitely a form of waste).

As an improvement, it was suggested the company change the production input sequence so as to avoid all unnecessary changeovers (i.e., to minimize replacement of semifinished products and materials). Figure 6-8 shows how these product types were organized under this new criteria.

Production prioritization (changeover factor)	Description	Problem
1	Products with specified dates (regardless of quantity)	• Frequent change-overs are required. • Semifinished products and materials must be switched whenever product types are switched. • Production input sequence is not finalized until the same day (production input sequence varies from day to day).
2	Product with the highest amount of orders (to be shipped on the same day)	
3	Product with the second highest amount of orders (to be shipped on the same day)	
4	Product with the third highest amount of orders (to be shipped on the same day)	
5	Product with the fourth highest amount of orders (to be shipped on the same day)	
6	Product with the fifth highest amount of orders (to be shipped on the same day)	

Figure 6-7. Changeover and Production Input Sequence Prior to Improvement (Including Problems)

Category	Criteria		Product (item) name					
Product	Date of manufacture is specified		A	B	C	D	E	F
			G	H	I	J		
	Date of manufacture is not specified		K	L	M	N	O	P
			Q	R	S	T	U	V
Semi-finished product	Semifinished products used in multiple products	Semifinished product W	A	B	C	D		
		Semifinished product X	E	F	G	H	I	J
		Semifinished product Y	T	U	V	O		
		Semifinished product Z	K	L	U	T		
	Semifinished products used in only one product		M	Q	R	S		
Materials	Materials used in multiple products		A	B	E	G	J	T
	Materials used in only one product		C	D	F	H	I	O
			K	L	M	N	U	V
			Q	R	S	T		

Figure 6-8. List of Use Criteria for Semifinished Products and Materials

Under, the category column in Figure 6-8, the first category, *Product*, shows a distinction between products that require a specified date of manufacture and products that do not require it. The second category, *Semifinished product*, shows a distinction between semifinished products that are used in multiple products and those that are used in only one product. A similar distinction is shown in the third category, *Materials*. As is shown in the figure, all product items except for item M, item Q, item R, and item S share the same semifinished products or materials with other items. In other words, if these items with common semifinished products and materials are strategically grouped together in the production sequence, the frequency of changeovers for secondary packaging processes can be greatly reduced. The next step was to create new product input sequence patterns for secondary packaging processes based on the information in Figure 6-8. The criteria used for these patterns are outlined in Figure 6-9.

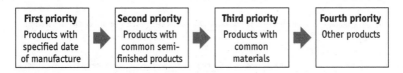

Figure 6-9. Prioritization Criteria of Product Input Sequence

As Figure 6-9 shows, products with a specified date of manufacture were given first priority in the input sequence for secondary packaging processes, then products with common semifinished products, and then products with common materials. This new order arrangement enabled the company to issue production specifications that minimized the number of required changeovers.

Improvement 3: Setting verification and reception time limits on previous-day order changes to level the production load. As was shown in Figure 6-3, before improvement was made the equipment operation time varied widely from day to day. Sometimes the production volume required for the day's shipments would be finished early, other times the factory had to run overtime to finish them. There were even occasions when working overtime was not enough, and some items went missing from shipments. The solution to this kind of problem is to level the production load (see Figure 6-10). Before actually trying to level the production load, several simulations were conducted, based on daily order information in order to check the feasibility of production load leveling.

The simulation method included coordinating the order information with various hypothetical postimprovement production conditions to verify whether or not it would be possible to level the production load. The simulation's conditions for the overall equipment efficiency and time-based operation rate are set as follows.

- The equipment load time is the regular shift time.
- Current values are used as the speed of secondary packaging processes for each product.
- The input sequence set for "Improvement 2" is set as the production sequence for each product, and the number of production runs and changeovers has been minimized.

Calculation of overall equipment efficiency rate for simulations

$$\text{Overall equipment efficiency rate} = \frac{\Sigma\,(\text{speed of secondary packaging processes for each product} \times \text{No. of orders for each product})}{\text{Equipment's load time (regular shift time)}}$$

$$\text{Time-based operation rate} = \frac{[\text{Equipment's load time (regular shift time)}] - [\Sigma\,(\text{changeover time} - \text{stoppage time due to equipment faults})]}{\text{Equipment's load time (regular shift time)}}$$

** Verification **

$$\text{Equipment's load time} = [\Sigma\,(\text{speed of secondary packaging processes for each product} \times \text{No. of orders for each product}) + \Sigma\,(\text{changeover time} - \text{stoppage time due to equipment faults})]$$

Figure 6-10. Overview of Production Load Leveling Verification

The results of these simulations verified that all orders could be handled by the production system without having to exceed the normal daily load time (regular shift time), that is, without having to resort to overtime operation of the production line. This means that production capacity exceeded the total number of orders (see Figure 6-5), so on days when the production capacity exceeds the day's number of received orders, you could use the extra production capacity to manufacture products for the product inventory, which helps to systematically maintain high efficiency. However, products that require a specified date of manufacture must be manufactured on that date and cannot be stored in the product inventory. Therefore, the production system must include a mechanism for dealing with order changes concerning products with a specified date of manufacture.

Consequently, the next step was to determine the approximate share of products with specified dates of manufacture and to determine whether or not they would preclude planning a production schedule that has systematically high efficiency. To determine this, we studied the order information and analyzed the relative shares of products with specified date of manufacture among the products with order changes (spot orders). The results of this analysis are shown in Figure 6-11.

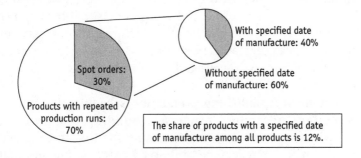

Figure 6-11. Share of Products with Specified Date of Manufacture among All Ordered Products

As the figure shows, our analysis indicated that only 12 percent of all ordered products (by volume) required a specified date of manufacture. Since all products other than these 12 percent do not require a specified date of manufacture, strategic amounts of these products can be maintained in a product inventory so as to enable a production schedule that has systematically high efficiency. In addition, in order to ensure that 100 percent of the products that do require a specified date of manufacture will be manufactured on the specified date, Miyagi Shimadaya's factory managers asked Shimadaya's head office to always submit the final order changes for the next day by 1:00 P.M. The reason for requesting the final order changes by 1:00 P.M. of the day before the manufacture date is that 1:00 P.M. is the latest time at which the production schedule for the primary packaging processes can be changed.

This timing was needed because, if the final order changes that arrive at 1:00 P.M. include orders for semifinished products that have not been completed by the primary packaging processes, there is still just enough time to manufacture those semifinished products before they will be needed at the secondary packaging processes the following day. If any of the order changes received on the previous day are for products that require a specified date of manufacture, those products will be the first to be supplied to the secondary packaging processes on the following day (i.e., at the start of the day). This is in keeping with the prioritization criterion that products that require a specified date of manufacture must have first priority in the input sequence for secondary packaging processes (see Figure 6-9).

Naturally, when the last order changes are received by 1:00 P.M. of the previous day, any of the ordered products that require a specified date of manufacture can be manufactured (if necessary) using the primary packaging processes after 1:00 P.M. on that same day (i.e., the day before the specified date of manufacture).

Improvement 4: Establishing production schedule planning steps for secondary packaging processes. Building upon what was learned while studying and implementing the first three improvements, the next improvement involved establishing production schedule planning steps for the secondary packaging processes, the goal being to plan production schedules that ensure the highest possible efficiency on a daily basis. The key to creating such production schedules is specifying a production load that makes full use of the production capacity each day, so that each day is a full day of production, with no early completion or overtime. The most important element is to establish steps whereby all of the spot orders (including those for products that require a specified date of manufacture) can be successfully incorporated into an ESP production planning system. After studying this matter, we decided to adopt the following approach for planning production schedules.

- When the weekly estimated orders arrive, the product items that have repeated production runs can be expected to require about 70 percent of the daily production capacity.
- Allocate these product items so that they will account for about 70 percent of the production capacity, starting with the products that past order information shows as having the highest number of repeated production runs.
- Allocate the quantity of each product item among these products so that they will be distributed evenly from day to day, based on average values taken from the weekly estimated orders.

- Allocate time (without specifying any particular product items) for the 30 percent of the products that will come from spot orders (last-minute order changes).

Using this approach, we devised the following ESP production schedule planning steps (see Figures 6-12 and 6-13.)

Planning of secondary packaging processes when weekly orders are received:

1. 30 percent of the load time is left empty to allow production time for spot orders.

2. The products from spot orders correspond to the 30 percent of the load time (on the shipment date) that is allocated for the spot orders (i.e., the products from spot orders are all manufactured on the date they are scheduled for shipment).

3. 70 percent of the load time is allocated to "prescheduled products" (products based on the weekly estimated orders). The volume is distributed evenly from day to day, based on average values taken from the weekly estimated orders.

4. After allocating the prescheduled products, if there is any operation time remaining within the 70 percent allocated to those products, the number of prescheduled products can be increased to take up the slack.

5. However, additional prescheduled product items are not manufactured as described in 4 above if the result would be a product inventory level of those items in excess of the upper limit.

6. If there is a lot of surplus operation time (a half day or more), a half day or full day of planned downtime can be scheduled.

Planning of secondary packaging processes when spot orders are received:

7. If spot orders account for more than the 30 percent share of load time allotted to them, any surplus load time from the pre-scheduled orders can be applied for production of spot order items.

8. If there is no surplus load time from the prescheduled orders, or if the surplus is not enough to cover the excess spot orders, production is cut back for product items that are already in the product inventory in order to accommodate all of the excess spot-order items.

Figure 6-12. Steps in ESP Production Schedule Planning

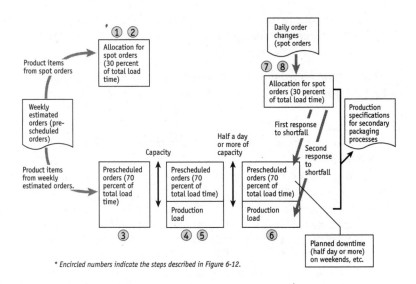

* Encircled numbers indicate the steps described in Figure 6-12.

Figure 6-13. Image of ESP Production Planning System

Improvement 5: Establishing a production schedule planning method for primary packaging processes. The planning method used to schedule production for primary packaging processes is based on the weekly production schedule for secondary packaging processes and includes estimates calculated from product inventory volume trend information. The following eight points describe the basic rules for planning production at primary packaging processes.

1. The production load for the primary packaging processes should be specified so that the production capacity is used fully (i.e., a full regular shift, without early completion or overtime).
2. Semifinished products that have a low turnover rate should be generally excluded from product inventory and should only be manufactured (via the primary packaging processes) on the same day or the day before being supplied to the secondary packaging processes.
3. Semifinished products that have a high turnover rate (generally, these are semifinished products used in prescheduled finished products) should be produced in quantities sufficient to maintain the specified inventory level.
   • Upper and lower inventory level limits are set for each of these semifinished products.
4. The amount of semifinished products having high turnover rates that will be produced is determined based on the weekly production schedule for secondary packaging processes.

5. Trend calculations using estimated weekly inventory amounts of semifinished products are made to help determine the current semifinished product inventory levels and the current production schedule for primary packaging processes.
6. Estimated inventory calculations are made and are compared with the specified limited inventory amounts for each type of semifinished product in order to determine whether or not additional semifinished products need to be manufactured.
   - Insertions are made in the production schedule when more semifinished products are needed to stay at or above the lower-limit inventory level.
   - Stops are made in the production schedule when fewer semifinished products are needed to stay at or below the upper-limit inventory level.
7. A production input sequence for primary packaging processes is established for each type of semifinished product.
8. Only one changeover per day should occur in the primary packaging processes.
   - To enable wide-variety, small-lot production, changeovers should occur only when absolutely necessary, but only one changeover should occur at primary packaging processes, if possible.
   - Although there are four production lines that are primary packaging processes, each line must not have more than one changeover per day.

---

1. Loss will occur whenever a changeover is performed in primary packaging processes.
2. Before, when semifinished products were manufactured in batches according to type, changeovers still occurred frequently.
3. This improvement enables changeovers in primary packaging processes to be reduced to just one per day.

Stage 4: Improving the Manufacturing Processes

Building on the improvement topics already addressed by TPM activities, improving your manufacturing processes stands as the fourth stage in establishing an ESP Production System. Miyagi Shimadaya launched some TPM (Total Productive Maintenance) activities in September, 1996 in order to boost the overall efficiency of its production equipment at various processes. Its subsequent decision to adopt ESP was viewed as an extension of these TPM activities. To make the most of ESP's effects, you must make improvements for the manufacturing processes that threaten to become bottlenecks when planning the ESP production schedule.

Bottleneck No. 1: Varying loads on specialized lines were a problem

Miyagi Shimadaya needed to change the three specialized lines with secondary packaging processes to nonspecialized lines. Their secondary packaging processes were laid out in three specialized production lines. Figure 6-14 illustrates this line configuration and layout.

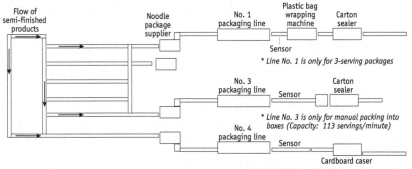

Figure 6-14. Line Layout

Before improvement, all three of these lines were specialized lines. Consequently, the load on each line varied depending on the number of orders received for each type of product—one specialized line might be overloaded, while another one was hardly being used. Figure 6-15 lists the distinguishing factors and conditions of three specialized lines. Line No. 1 was excluded from the improvement to unify these specialized lines because it packaged a different product type (different number of servings).

Improvement 1: Unifying the width of conveyor guides (unification item 1 for Line No. 3 and Line No. 4). In Line No. 3 and Line No. 4, different fixed-width settings were used on the conveyor guides that carried the soup packages to be added to the noodle packages. Before improvement, these fixed-width conveyor guides could not be adjusted. If the production planners tried sending the narrower soup packages along the conveyor guide whose width was fixed for the wider soup package, the packages would shift off cen-

	Product type	Semifinished product	Added product/ material	Final product packaging
Line No. 1	Three-serving package		Specialized	Manual packing into boxes
Line No. 3	One-serving package	Common	Partially specialized	Manual packing into boxes
Line No. 4	One-serving package		Partially specialized	Autopacking using wraparound cases

Figure 6-15. Distinguishing Factors and Conditions of Specialized Lines

ter and cause problems at subsequent processes. That is why the production planners decided to use a separate specialized production line for each of the two soup package sizes.

The ESP improvement group started by investigating why the narrow soup packages were shifting around on the wide conveyor. This investigation revealed that when the narrow soup packages traveled along the wide conveyor, they got twisted diagonally and then got caught on the conveyors. This was due to the packages' resistance to the conveyor's guide rail. The ESP improvement group then decided to try to develop a resistance-free conveyor guide rail. After trying out several different conveyor guide rail designs, they devised a wide conveyor guide that worked for conveying narrow soup packages, which meant that they were now able to use the same conveyor guide width for both the No. 3 and No. 4 lines.

Improvement 2: Unifying the packaging type (unification item 2 of specialized parts of Line No. 3 and Line No. 4). On both Line No. 3 and Line No. 4, the final packages were a case that held 10 servings and a larger box into which three of these cases were packed. The major difference between the two lines was that the cases were packed manually on Line No. 3 and automatically on Line No. 4. The automated packaging device (wraparound caser) on Line No. 4 could only work with a certain size case, so any other size had to be sent to Line No. 3 for manual packing.

The ESP improvement group did a cost analysis of the packaging processes in the two lines, and determined that cost benefits could be gained by unifying the two processes. The target factors for this cost analysis included:

- Equipment remodeling cost and running cost for revising specifications to enable all products to be manufactured on both lines.

- In addition to the costs in the previous bullet item, equipment investment cost and running cost for installing an autopackaging machine on line 3.
- Comparison of autopackaging machine costs to labor costs for manual packing. This is assuming that the autopackaging machine will require human supervision.
- Comparison of materials costs on both lines.

The results of this analysis led the improvement group to decide that the most cost-effective improvement would be to make Line No. 4 like Line No. 3, which is a more adaptable line.

Effects of improvements in secondary packaging processes made at Stages 1 through 4

As a result of the improvements made at Stages 1 through 4 in establishing an ESP Production System, the hours of operation for the secondary packaging processes were improved (see Figure 6-16). Clearly, the improved production lines helped to minimize waste while maximizing production efficiency.

		Regular operating hours	Overtime hours
Before improvement	Monday		
	Tuesday		
	Wednesday		
	Thursday		
	Friday		
	Saturday		
	Sunday		
After improvement	Monday		
	Tuesday		
	Wednesday		
	Thursday		
	Friday	Operations end early on Friday	
	Saturday	due to planned stoppage	
	Sunday		

Figure 6-16. Hours of Operation for Secondary Packaging Processes, Before and After Improvement

As Figure 6-16 shows, when the production planners in the improvement group at Miyagi Shimadaya put together their weekly production schedule (in Improvement 4 at Stage 3—on page 196), they were able to schedule an early end to operations each Friday Having this planned stoppage each Friday gave the company room to respond to last-minute order changes, requests for additional products, and so on. In other words, since their daily operation hours had become stabilized, they were able to handle all types of orders within their regular weekly operation hours, plus they had additional planned stoppage-hours available for making adjustments when needed.

Bottleneck 2: Setup/removal times at primary packaging processes were a major loss factor

Figure 6-17 illustrates the loss factor that was found when the improvement group looked for bottlenecks among its primary packaging processes. To help achieve maximum production efficiency by enabling full usage of regular operation hours (without early stops or overtime) every day at the primary packaging processes, Miyagi Shimadaya decided it was essential that loss due to setup/removal times be eliminated, since setup/removal times accounted for 71 percent of total loss.

Figure 6-17. Loss Ratios of Setup/Removal Times at Primary Packaging Processes

Improvement 1: Reduction of setup time. Setup time accounted for 25 percent of total loss. This was addressed with improvements made via the following steps.

1. The Miyagi Shimadaya improvement team analyzed setup operations and categorized them into two types, which are operations that have to be done on the same day, and operations that do not have to be done on the same day.

2. The improvement team combined all operations that did not have to be done on the same day with the previous day's removal operations.
3. Next, the improvement team studied whether any of the manual setup operations could be automated. They automated the equipment warmup operation by installing an autostop timer.
4. The improvement team studied various alternative sequences of manual setup operations in order to minimize idle (standby) time. They then standardized the best alternative setup operations by specifying them in an operations manual.

Note: the above improvements changed the total setup time from 62 minutes to 47 minutes, a 15-minute reduction.

Improvement 2: Reduction of removal (cleaning) time. Removal (cleaning) time had accounted for 46 percent of total loss, so Miyagi Shimadaya implemented loss-reducing improvements using the following steps.

1. They mechanized some parts of the removal operations.
2. They modified the equipment to reduce the standby time involved in setup operations.
3. They used two people instead of one for removal operations, thereby reducing the removal time.
4. They standardized the optimum removal procedures (those with the smallest loss).

Note: the above improvements changed the total removal time from 85 minutes to 41 minutes, a 44-minute reduction.

BENEFITS OF MIYAGI SHIMADAYA'S ESP PRODUCTION SYSTEM

By adopting an ESP Production System, Miyagi Shimadaya was able to build a new production management system that greatly improved their equipment utilization rate. Figures 6-18 and 6-19 illustrate various benefits gained by implementing ESP, which included boosting the overall equipment efficiency rate into the 90 percent range. They also realized substantial improvements in their time-based and labor-based production output.

These improvements enabled Miyagi Shimadaya to reduce its labor costs, as well as its product costs. The company also gained the ability to avoid major variation in production schedules (production specifications), and this freed up time they could use to schedule regular small-group improvement activities, thereby bol-

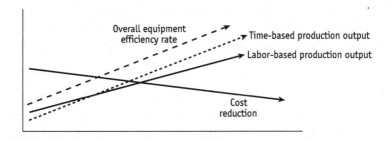

Figure 6-18. Effects of ESP Production System

stering such activities. Figure 6-19 also shows, in detail, that the improvement in the overall equipment efficiency rate was the result of improvements that reduced various types of time loss, which allowed Miyagi Shimadaya to make more effective use of the time as production time. Production output was increased most dramatically in terms of labor-based output. Again, this was the direct result of reducing time loss (standby time, early completion, etc.). This case study stands as an excellent example of how important it

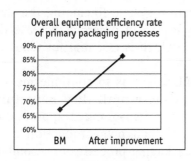

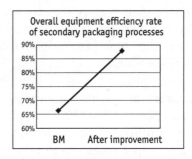

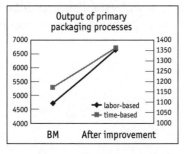

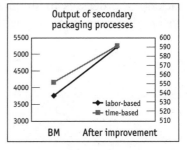

(The unit of output is one serving. The left scale is the number of servings per production line operation hour and the right scale is the number of servings per labor hour.)

Figure 6-19. Benefits of ESP Production System

is for process-industry companies, such as Miyagi Shimadaya, to undertake these types of activities to build a production system in which production equipment is used much more efficiently.

Issues for the Future

The improvement activities described in this case study were indeed successful, but not enough to satisfy the people at Miyagi Shimadaya. One remaining issue was the existence of inventory at the primary packaging processes and secondary packaging processes. And, as inventory is something that has to be managed, their next frontier for improvements will be reducing inventory, which they consider a form of loss.

As part of this inventory-minimization effort, they would like to fully synchronize production at both the primary packaging processes and secondary packaging processes. Their first order of business will be to reduce the retention time of inventory located between the primary packaging processes and the secondary packaging processes. They believe it will be necessary to fundamentally revise their factory's entire production system (including the shift system and the regular operation hours) in order to increase the operation time per equipment unit, which is essential for reducing retention time.

The ultimate goal at Miyagi Shimadaya is to continue making ambitious improvements such as these until they have fully matched the ESP Production System's two-pronged approach of synchronization and equalization to their own production cycle (from order reception to primary packaging processes to secondary packaging processes to shipment). This will establish a production system that runs smoothly, without a single breakdown or obstacle.

Case Study 2:
Izumi Industries and Okegawa Plant

A pile of inventory means a heap of trouble!

Izumi Industries was founded in 1923 by Toukichi Izumi. At that time, there were very few automobiles in Japan, and the few that existed were foreign-made and extremely expensive. Izumi Industries' founder, Toukichi Izumi, laid the foundations of today's Izumi Industries by working in the automotive repair business, where he rebored cylinders for rebuilt engines and made oversized pistons to fit the rebored cylinders. Today, Izumi Industries is located in Kawagoe and has two plants, Okegawa Plant and Tsuruoka Plant, where the main products include automobile pistons and cylinder liners.

The main type of product at the Okegawa Plant is aluminum pistons for internal combustion engines. This plant boasts an integrated manufacturing system that includes everything from melting and casting the metal to finishing processes. Pistons are a central part of any engine, whether it's for a car, truck, bus, boat, or construction machine. It is a part that plays a key role in converting energy from fuel combustion to drive power, so there are strict requirements for this part's characteristics, including a need for micron-level precision on nearly all of its outer surfaces. In addition, other factors that require increased precision have come into play in recent years, including the need for higher output, regulations on gas emissions, fuel economy, and noise levels, and competition to extend product life.

PROBLEMS BEFORE INTRODUCING THE ESP PRODUCTION SYSTEM

The Okegawa Plant had already implemented TPM (Total Productive Maintenance), a type of companywide activity aimed at minimizing equipment-related loss and boosting productivity, and they considered implementing the ESP Production System as an extension of these TPM activities. As indicated by the list in Figure 7-1, the Okegawa Plant struggled with a number of problems prior to introducing ESP. For some of these problems, the company had to resolve production management issues before much success could come from TPM activities. In fact, the improvements that could be made through TPM were likely to slip back to their original conditions unless changes in production management were made to support the improvements. This is what led the company to decide that building an ESP Production System as a means of fundamentally improving its production management system could provide just the type of support needed for its TPM activities.

Issues Related to Production Management

The Okegawa Plant focused on seven issues related to production management.

1. Reducing in-process inventory and product inventory.
2. Boosting productivity.
3. Leveling the production load.
4. Achieving zero late deliveries.
5. Reducing production planning labor hours.
6. Getting more precise data for planning standards.
7. Getting more precise data on production results.

The company wanted to reduce overall costs by pursuing these issues. As Figure 7-2 shows, ESP's Six Guarantees address all seven of the production management issues. When the company recognized this, they decided to adopt the ESP approach.

Before adopting ESP, the Okegawa Plant carried out production scheduling on a monthly basis, and the planning tasks involved a lot of manual work. It was not well coordinated with the weekly cycle of order reception, so their production load was based on the number of received orders. Consequently, they were not able to achieve a high level of production efficiency. Production planning

Problems Before Introducing the ESP Production System

- The casting schedule and processing schedule were not being drafted as linked schedules, and this tended to create parts depots (in-process inventory).

- The casting schedule tended to be extra busy (with overtime) during the first half of the month and slow during the second half.

- Planning standards were set based on experience, and so there was often a poor balance between process capacity and production load.

- There was a wide variation in production load, due in part to a lack of interchangeability and flexibility among the production lines.

- The equipment's overall operation rate was low and production was often behind schedule.

- Even though kanban were being circulated, required products came up short while inventory of unneeded products accumulated.

- The plant was slow in adopting office automation technologies for its clerical tasks, which exacerbated problems such as information missing from forms, clerical delays, and inaccurate information on production results, inventory levels, etc.

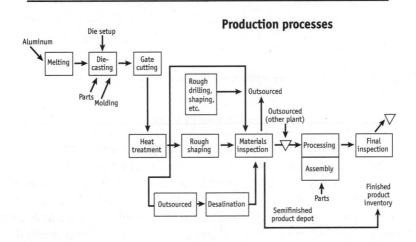

Production processes

Figure 7-1. Main Production Processes and Problems Before ESP

was done by veteran employees who relied on their own experiences, and so planning methods varied from one person to the next, and anyone outside of the group of planners would have had a hard time understanding the planning standards they used. They also made it a top priority to introduce an EDP (Electronic Data Processing) system to computerize production management tasks (something they had wanted to do for a long time) and they had just launched some preliminary activities toward that end.

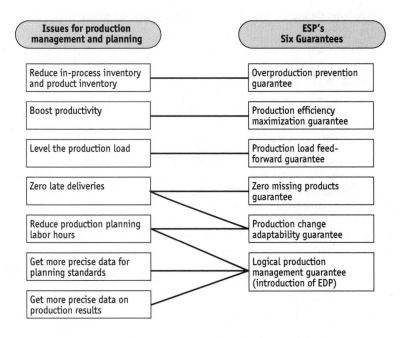

Figure 7-2. Production Management Issues and ESP's
Six Guarantees

Role of ESP and TPM at the Okegawa Plant

The role (position) of the ESP Production System in the Okegawa Plant's program of TPM activities was to create a new production management system that would help maximize the benefits gained through TPM-style improvement activities. In addition, improvement themes were to be selected from a production management perspective and the making of improvements would be promoted by assigning various improvement topics to all relevant departments in the company. In particular, the shortening of lead time would be the central focus of many improvement themes, since it was an important contributor toward reducing inventory.

The Okegawa Plant's TPM activities were mainly focused on cost-cutting activities to support the company's overall goal (called the "CC30 Plan") of cutting costs 30 percent. Naturally, ESP was introduced as part of these cost-cutting efforts (see Figure 7-3).

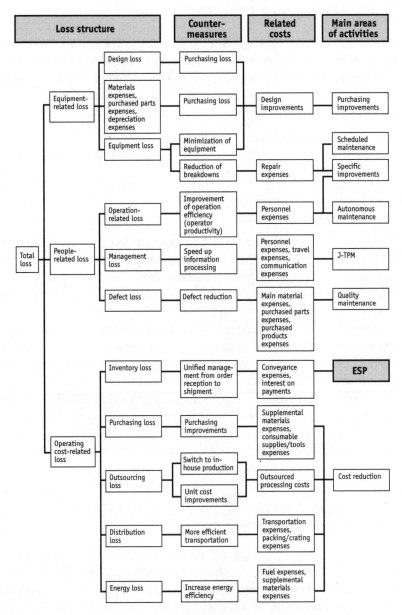

Figure 7-3. Relationships between Target Costs and Main Areas of Activities

Targeted Values and Goals

The main goals (progress toward these goals serve as a criterion for evaluating these activities) were reducing the volume of product inventory and in-process inventory and reducing the labor hours involved in production scheduling. The target values set for these goals were as follows.

- Reduction in volume of product
  inventory: 60 percent reduction
- Reduction in volume of in-process
  inventory: 60 percent reduction
- Reduction in labor hours involved
  in production scheduling: 50 percent reduction

Although the ultimate goal in reducing product inventory and in-process inventory is the achievement of zero inventory, you cannot accomplish it until your lead time from purchasing of parts to manufacturing becomes shorter than the lead time for meeting customer requirements, and the manufacturing capacity becomes greater than the load imposed by received orders. While striving to improve the production system so that inventory is reduced to as close to zero as possible, you need some level of inventory to enable prompt and flexible responses at peak load periods. ESP techniques are used to establish production management methods whereby customer orders are separated from factory-floor operations for the sake of maximizing production efficiency. Therefore, you set the target inventory levels according to the levels you can achieve through production management improvements.

OVERALL CONFIGURATION OF ESP AND MAIN AREAS OF IMPROVEMENT IN DATA PROCESSING

Figure 7-4 includes a diagram of the overall configuration of the ESP Production System. It shows how information about various functions has been stored in digital form to speed up the processing of relevant data. In this section we will discuss the Okegawa Plant's main areas of improvement in regard to this.

1. Expanded EDP in sales planning.
   - Shorten information-related lead time and labor hours by using online information from customers and using more EDP applications for sales planning and information-gathering tasks.

2. Expanded EDP in production scheduling.
    - Establish automatic generation of monthly schedules.
    - Establish full deployment (processing, casting, parts, etc.) of each schedule and interlinking of schedules to reduce in-process inventory.
    - After using confirmed information from your customers to check for missing items, create the confirmed schedule for the next ten days.
    - Study current changeover efficiency and devise ways to improve production efficiency.
3. Expanded EDP in collecting and processing production specifications and production results statistics.
    - Establish automatic generation of production specifications.
    - Use bar-code tags to issue production specifications.
    - Use bar codes to reduce labor in entering production results data.
    - Seek to establish real-time monitoring of results in order to provide more accurate information for tasks related to inventory control, cost estimations, etc.
4. Expanded EDP in shipping management tasks.
    - Use bar codes to reduce labor in entering shipping data.
    - Establish real-time processing and monitoring of shipping results.
    - Link production results data with autowarehousing functions to provide more accurate information for shipping management and inventory control while reducing labor requirements.

Flow of Scheduling Processes

Figure 7-5 on page 215 shows a flowchart of the processes when creating linked schedules based on information from sales plans. The basic logic of ESP is incorporated into the various scheduling functions in order to implement the Six Guarantees. A further description of these scheduling functions as they relate to Izumi Industries—which operates the Okegawa Plant—is discussed next.

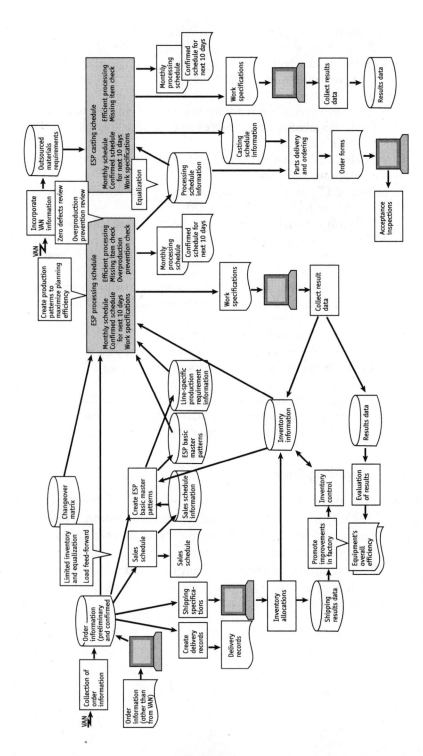

Figure 7-4. Conceptual Diagram of ESP Production System (Overview)

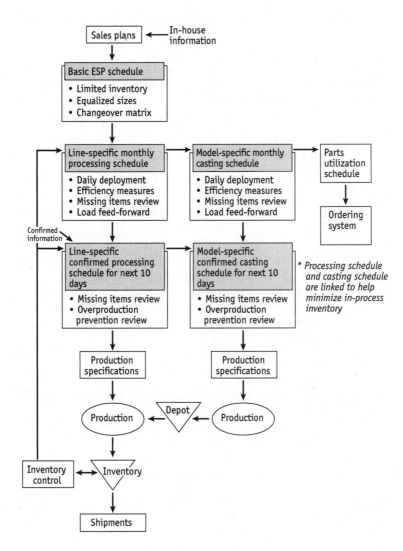

Figure 7-5. Flowchart of Scheduling Processes

Scheduling Methods and Characteristics at Izumi Industries

We will now look at the methods Izumi Industries used to determine equalized sizes.

1. Establish number of production shifts based on information from sales plans.
   - Production volume per shift is equal to operating time per shift/CT (cycle time)

- Average inventory allocation per day is equal to volume in sales plan/No. of working days

---

1. When volume in sales plans for N months < Volume in sales plan [N + 1]: Required production volume = (Volume in sales plans for N months + Volume in sales plan [N + 1]) / 2.

   Leveled load, determined based on a two-month average load

2. When volume in sales plans for N months > Volume in sales plan [N + 1]: Required production volume = Volume in sales plans for N months.

- Number of production shifts as determined the value of X, below.

(Standard labor hours (h) x 2-shift operations) / average volume allocated to inventory = X.

If X =:	3 or less	5 shifts	91 or less	1/2 shift
	3 to 7	3 shifts	Other (S rank)	5 shifts
	8 to 30	2 shifts	E rank	1/4 shift
	31 to 90	1 shift		

---

2. Products with high volume in sales plans are given larger equalized sizes.
   - These methods give production efficiency high priority, while minimizing the number of production runs.

Methods for Determining Limited Inventory at Izumi Industries

The methods used at Izumi Industries to determine limited inventory are described below.

- Limited inventory level is equal to the average daily volume allocated to inventory based on the sales plan—3 days.
- Based on the timing by which they receive confirmed order information, they set the period of the confirmed schedule as the next 10 days. If there are three days worth of average volume allocated to inventory, having at least one production run per 10-day production schedule should suffice to prevent missing items.
- Since processing for higher efficiency is performed within 10-day periods, they determined that three days worth of average volume allocated to inventory would be necessary.
- They devised a method for reducing inventory levels of stale products.

Planning Flow for Processing Schedule at Izumi Industries

The required production volume at Izumi Industries was calculated based on factors such as the sales plans, finished product inventory levels, in-process inventory levels at inspection processes, volume of workpieces scheduled for processing, volume of undelivered goods, volume of goods scheduled for delivery, and limited inventory specifications. Next, the production volume was worked into in a daily schedule that was processed to fix volume peaks and to improve efficiency, so that a monthly processing schedule could be generated automatically. Figure 7-6 shows a flowchart of these planning processes.

Although the confirmed schedule is a 10-day schedule based on the monthly processing schedule, you can use the monthly schedule for efficiency-boosting measures as long as missing items do not occur and the daily schedule does not change.

To plan the flow for a confirmed 10-day processing schedule, you must first perform a missing item check on the monthly processing schedule, based on confirmed customer orders (equal to confirmed inventory allocation amounts). You then adjust for any missing items that are found and then confirm the 10-day processing schedule. Figure 7-7 on page 219 shows a planning flowchart for the confirmed 10-day processing schedule.

Deploying the Daily Processing Schedule and Efficiency- Boosting Measures

The logic behind auto-allocation of the daily processing schedule and efficiency-boosting measures is explained in Figure 7-8 on page 218. The basic approach uses a changeover matrix to determine the optimum input sequence that will provide for maximum production efficiency within the range in which missing items will be prevented.

Planning Flow for Casting Schedule for the Okegawa Plant at Izumi Industries

The processing schedule information is used to deploy product item numbers (including the casting volume) in the casting schedule, inventory is allocated as semifinished product inventory and in-process inventory for casting, then the remaining allocations

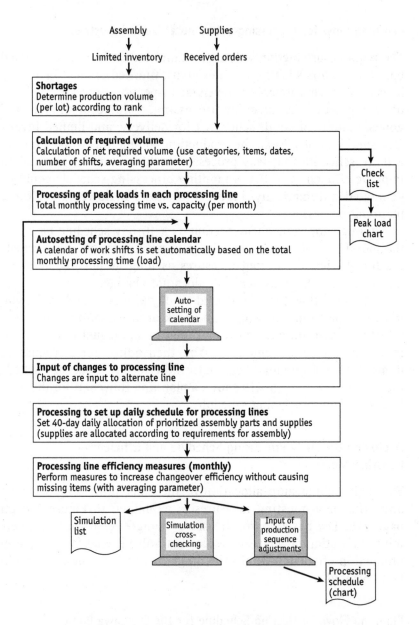

Figure 7-6. Flowchart of Processes for Automatic Generation of Monthly Schedules

are made for the casting specifications and the casting schedule is worked out. Using the materials requirement information received from outsourced processing companies and other affili-

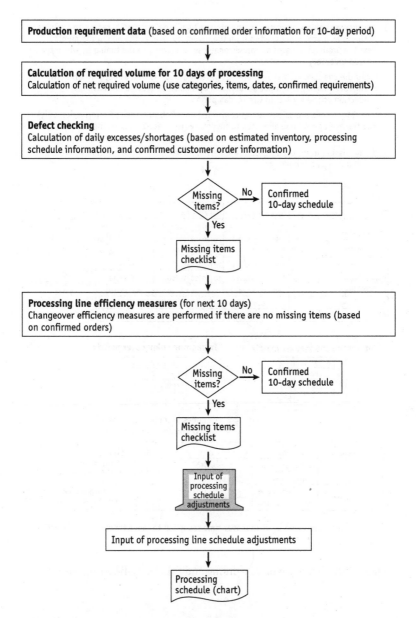

Figure 7-7. Planning Flowchart for Confirmed 10-day Processing Schedule

ated companies, the net volume for the casting schedule is calculated and deployed in a daily schedule, after which the casting schedule is generated automatically. One basic requirement for

1. The average volume allocated to inventory, which is determined based on the volume in the sales plans, is subtracted from the inventory level at the end of the month. Next the production dates are provisionally set according to the timing by which the limited inventory is allocated.

2. Using the changeover matrix shared by several groups as well as each group's own changeover matrix, the input sequence is optimized to provide for maximum production efficiency within the 10-day period.

3. Inventory trends are calculated as confirmed information (received information only) and the average volume allocated to inventory (from No. 1 above), then a missing items check is performed.

4. If any missing items occur, groups are moved forward, up to the position where the missing items no longer occur.

Line A

Product item number	Daily schedule	1	2	3	4	5	6	7	8	9	10	11	
A1	Production volume		50		50					50		50	Input sequence
Inventory at end of month 100	Amount allocated	20	20	20	20	20				20	20	20	20	B1→A1 →B1 →A1
Limited inventory level 80														→A2→B1 →A1 →C1
Equalized size 50	Inventory trend	80	110	90	120	100				80	110	90	120	
B1	Production volume	30		30	30					30	30			
Inventory at end of month 50	Amount allocated	10	10	10	10	10				10	10	10		Total changeover time
Limited inventory level 50														50 + 50 + 50 + 10
Equalized size 50	Inventory trend	70	60	80	70	60				80	70	60		+ 50 + 50 + 60 = 320
A2	Production volume					45								
C1	Production volume										70			

Intergroup changeover matrix

	A	B	C	D
A		50	60	65
B	50		70	60
C	60	70		50

Intragroup changeover matrix

	A1	A2	A3	A4
A1		10	15	10
A2	10		20	15
A3	15	20		15

Changes in input sequence based on changeover matrices

Product item number	Daily schedule	1	2	3	4	5	6	7	8	9	10	Input sequence
A1	Production volume				50	50			50			B1→B1 →B1 →A1
B1	Production volume	30	30	30								→A1→A1 →A1 →A2 →C1
A2	Production volume									45		Total changeover time
C1	Production volume										70	0 + 0 + 50 + 0 + 0 + 10 + 60 = 120

Use the time saved by changeover time reduction to move forward the production schedule for the next 10 days.

Figure 7-8. Overview of Scheduling and Efficiency-Boosting Measures

this processing is that the unit used in the automatically generated casting schedule is the cast product model group, casting method, and equipment set (paired casting equipment units). The daily schedule for each cast workpiece unit is worked out in accordance with the master guide for setting requirements. Figure 7-9 shows a flowchart of these processing steps. The information

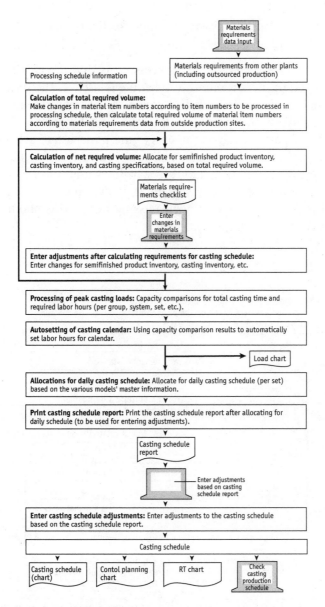

Figure 7-9. Flowchart for Autogeneration of Casting Schedule

in the planning flow for the 10-day casting schedule (in Figure 7-7) differs only in its period (10 days instead of a month). Otherwise, it has the same planning flow as is shown in Figure 7-9.

Planning Flow for Parts Usage Schedule for the Okegawa Plant at Izumi Industries

The basic concept behind this processing is that the casting schedule that you draft as a monthly schedule is used as the basic schedule, which is then developed in terms of the required amounts of parts based on the parts table. Once you have calculated the amounts of parts required for usage, the parts usage schedule can be generated automatically. In addition, you can display parts usage conditions by comparing monthly schedules and ten-day schedules when planning the ten-day schedules, which allows for the timing needed to anticipate and prevent missing items. This planning flow is illustrated in Figure 7-10.

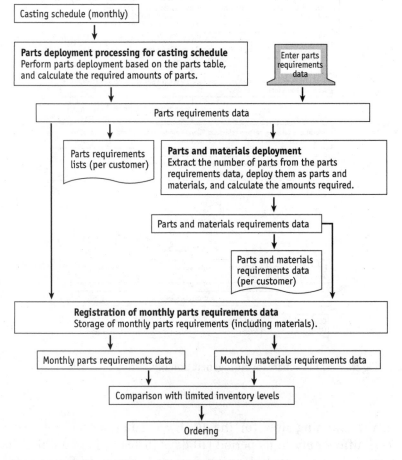

Figure 7-10. Planning Flow for Parts Usage Schedule

Comparison of Production Planning Steps

The following differences exist between production scheduling, based on the Okegawa Plant at Izumi Industries previous methods, and synchronized and equalized scheduling, based on EDP technology. As is shown in Figure 7-11, the lead time for creating a schedule is 11 days less in the latter case.

- **Planning based on previous (manual) methods.** Draft processing schedule based on sales plans.
  - Use proposed casting schedule to issue responses concerning materials supply dates.
  - Revise processing schedule according to materials supply conditions.
  - Use inventory stocking responses as basis for adjusting sales staff assignments and amounts to be delivered.
  - Too much time is needed to determine the final schedule, due to numerous adjustments.
- **Synchronized and equalized scheduling (using EDP).** All schedules from processing to transfer of control are created using EDP processing based on sales plans, so the lead time for scheduling can be reduced. Also, since operations are based on ten-day schedules, schedules can be created based on the latest confirmed order information.

RESULTS OF ACTIVITIES AT IZUMI INDUSTRIES

For Izumi Industries, the achievement of their long-pursued goal of introducing EDP technology for production scheduling tasks brought about many positive effects, including the following.

- Establishing a synchronized and equalized production management system that covers everything from receiving orders to shipping products.
- Switching from monthly schedules to ten-day schedules.
- Devising more efficient production input sequences.
- Reducing the lead time and labor hours required for production scheduling.
- Making production schedules more adaptable to spot orders.
- Exercising better management of in-process inventory and product inventory.

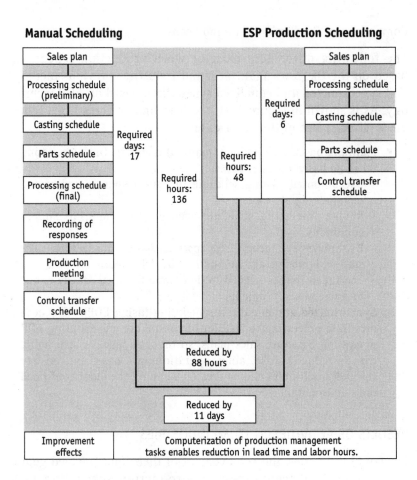

Figure 7-11. Comparison of Production Planning Steps

As a result of these and other improvements, the Okegawa Plant's overall operation rate for equipment was raised beyond 85 percent and, best of all, the company was awarded Japan's PM Prize (currently renamed as the TPM Excellence Prize). Figure 7-12 shows trend graphs for in-process inventory and product inventory that illustrate the steady success achieved through these activities.

In-process Inventory Trend Graph

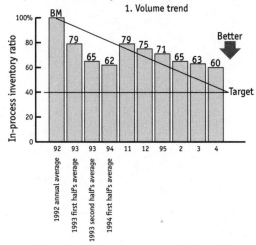

Product Inventory Trend Graph

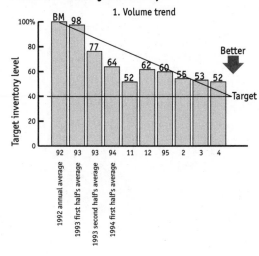

Figure 7-12. In-process Inventory and Product Inventory Trend Graphs

8

Case Study 3:
Tokyo Chuzosho, Gunma Plant

From Red Ink to Black

Since its founding in 1950, Tokyo Chuzosho has been a leading specialized manufacturer in Japan's foundry industry, with pig iron casting operations centered at its Gunma Plant and light alloy casting operations at its Saitama Plant. Tokyo Chuzosho has grown over the years as a manufacturer of products for the automotive, construction machinery, and shipbuilding industries. However, since 1980, the dual forces of industrial restructuring and technological innovation helped to create trends toward smaller and lighter products and wide-variety, small-lot production, and Tokyo Chuzosho was directly affected by both of these trends. In fact, this double whammy dealt a major blow to industrial materials suppliers in general and, in particular, to the pig iron casting operations at Tokyo Chuzosho's Gunma Plant. An additional factor behind these changes in the industry was the set of structural problems that beset Japan's foundry companies at that time.

Tokyo Chuzosho sought to extricate itself from this difficult situation by fundamentally changing its organization. Its new strategy for survival, despite the gloomy outlook for the foundry industry, was to launch TPM (Total Productive Maintenance) activities starting in January 1986. Its successes in this new strategy were so great that Tokyo Chuzosho was awarded Japan's PM Excellent Industries prize for the 1988 business year. As part of those TPM activities, Tokyo Chuzosho also sought to fundamentally revise its production management system by introducing ESP to move toward a synchronized and equalized production system. They worked hard to improve manufacturing processes, and the results led directly to the accomplishment of all of the company's business goals. The following describes how Tokyo Chuzosho's Gunma Plant overcame its harsh business environment and achieved remarkable success by developing an ESP Production System.

BACKGROUND TO DEVELOPING ESP AT THE GUNMA PLANT

At Tokyo Chuzosho's Gunma Plant, the setup was that of a typical manufacturing plant, which means that a lot of costs were determined by how well the existing equipment was being used (i.e., whether standard cycle times were short and whether the equipment's overall operation rate was being maintained at 90 percent or above). Consequently, their initial efforts after launching TPM activities were centered on shortening the standard cycle time for production equipment and minimizing the six losses related to production equipment by improving manufacturing processes. (The six big losses of machine ineffectiveness are 1. breakdowns, 2. setup and adjustment loss, 3. idling and minor stoppages, 4. reduced speed, 5. defects and rework, and 6. startup and yield loss.) Meanwhile, the Gunma Plant was a typical wide-variety, small-lot production facility, serving over 30 different companies and handling over 2,000 product models (items).

Realizing that they could not become more profitable by continuing with the current usage of equipment (aimed at short standard cycle times and an equipment's overall operation rate of 90 percent or above), the Gunma Plant's managers decided that they needed to truly revolutionize their company by creating a new production management system. They focused on two areas: reception of orders and production management problems.

1. *Reception of orders at the Gunma Plant.* The system for receiving orders at Tokyo Chuzosho's Gunma Plant is part of the plant's wide-variety, small-lot production system and therefore entails a lot of variation. As was mentioned above, the Gunma Plant's customers include over 30 different companies and the plant receives orders for over 2,000 different product items. The Gunma Plant uses kanban to issue delivery specifications for each customer, and thus the distribution of kanban makes up a sizeable part of the plant's operations. In addition, the plant must

deal with variations in orders due to last-minute rush orders and cancellations.

2. *Production management problems at the Gunma Plant.* In view of the variety of problems being faced, the production management system at Tokyo Chuzosho's Gunma Plant was rather weak despite the healthy level of orders being received. The company's practice had been to draft a monthly production schedule only once, at the start of each month. Production fell behind schedule so often that they became used to it, and nothing much was done about products that were lagging behind in the production system. As a result, the plant was unable to flexibly respond to last-minute order changes, and people simply did what they could on an ad hoc basis. In other words, there was a lot of doing without much planning.

The company was not able to completely wean itself from the mass production model that seeks economy of scale in manufacturing, and, consequently, they were unable to fully respond to customer needs for wide-variety, small-lot production. The end result was that items went missing from some shipments even while inventory levels rose, last-minute changes in the production schedule caused confusion at in-house manufacturing processes, and the production system was beset with waste due to various factors, including unreasonable requirements, inconsistent rules, and wasteful procedures. Planning a production schedule requires certain kinds of expertise and experience, and at the Gunma Plant the job of drafting production schedules was entrusted to the department manager. The existing production system was clearly in distress, and it was high time for the ESP Production System to come to the rescue.

GENERAL CONCEPT BEHIND ESP AT TOKYO CHUZOSHO

The general concept behind developing the ESP Production System at Tokyo Chuzosho was to fully apply *all* of its basic principles, which are described in more detail elsewhere in this book. So we will only briefly outline them below. ESP is premised upon:

- Maintaining zero missing products (including zero late deliveries) as the primary evidence of its reliability, *while*
- Working to maximize production efficiency, *by*
- Effectively using limited inventory (controlled minimum inventory), *by*
- Using equalized units for product item numbers in small lots, *by*

- Carrying out synchronized and equalized production, *which*
- Helps to minimize variation in the production schedule's daily production yield and enables flexible responses to rush orders and canceled orders.

Methods for achieving these objects include:

- Using computers.
- Establishing production management logic early on, *thereby*
- Weaning the company away from a production management system that is dependent upon an inner circle of managers.

The basic concepts behind ESP as it was implemented at Tokyo Chuzosho are outlined in Figure 8-1.

Figure 8-1. Basic Concepts behind ESP at Tokyo Chuzosho

Synchronization and Equalization at the Gunma Plant

There were two aspects of synchronization at the Gunma Plant. The first was synchronization in terms of synchronizing delivery with the customer's delivery deadline. The need for this type of synchronization goes without saying. Missing items occur when this aspect of synchronization fails. The second aspect of synchronization was synchronization between in-house upstream processes (such as the cores, molten metal, sand, and main mold processes, which should be considered as corresponding to the parts-and-materials supply processes in most types of manufacturing) and the metal casting production line. To devise this synchronization, Tokyo Chuzosho made use of limited inventory and in-house kanban methods with a view toward establishing a flexible production system.

There were also two aspects of equalization at the Gunma Plant. The first was the equalization of production lot sizes (lot sizes specified for each product model). This meant that each product item in the production flow had a predetermined lot size that was always used. Thus, production lot sizes were standardized and stabilized for each product item at the Gunma Plant. For example, if the lot size for product item A is 30, then 30 is always the number of units indicated in the production specifications issued for that product item. Because the lot size was predetermined for each product item, the production specifications for each product item needed only to contain a Go or Stop signal, which greatly simplifies volume management.

The second aspect of equalization was equalization of supplied materials (cores and molten metal). Since the production lot size of each product item had been equalized, the parts and materials (cores and molten metal) required for each product item could always be supplied in equalized amounts, so there was no longer any need to struggle with the difficult task of managing the volume of supplied parts and materials. The areas of synchronization and equalization at Tokyo Chuzosho's Gunma Plant are outlined in Figure 8-2.

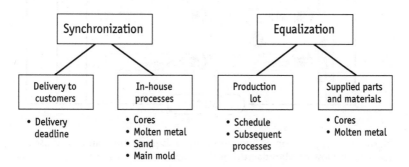

Figure 8-2. Areas of Synchronization and Equalization at the Gunma Plant

Figure 8-3 shows how synchronization, equalization, and in-house kanban are applied in the overall flow of the Gunma Plant's manufacturing processes.

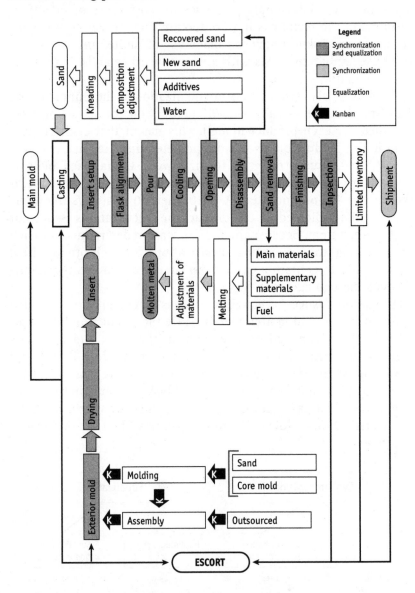

Figure 8-3. Application of Synchronization, Equalization, and In-house Kanban in Flow of Manufacturing Processes

Principles of Limited Inventory and Production Specifications at the Gunma Plant of Tokyo Chuzosho

The amount of limited inventory depends on the schedule planning period (in this case, ten days). In other words, in order to ensure that the limited inventory is adequate for providing the number of products required for synchronization with the customer's delivery deadline, the company calculated that a ten-day inventory of products would suffice, and therefore production was scheduled so as to maintain a ten-day inventory of certain products.

The product items for which a ten-day inventory is maintained are those that are constantly in the production flow, and the production load for such items is fed forward, based on the lead time. There is no limited inventory of spot-ordered product items. Instead, production is fed forward based on the lead time plus ten days, so that nothing will prevent the factory from meeting the deadline for spot-ordered product items within the ten-day schedule planning period.

A detailed description of the synchronization and equalization of production lots will be provided in this chapter, but let us note now that the amount of each production lot remains constant regardless of the size of the received orders. If the predetermined lot size is 30 units, and the received orders average 60 units per day, then two lots of 30 units each will be manufactured each day (rather than one lot of 60 units). Figure 8-4 shows a simple illustration of the relationship among limited inventory, received orders, and production lots.

Limited inventory: 300 units
Production lot size: 30 units

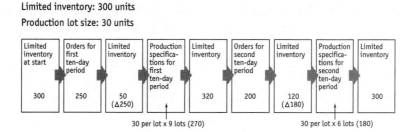

Figure 8-4. Relationship among Limited Inventory, Received Orders, and Production Lots

Restrictions on Production Scheduling at the Gunma Plant

The ESP Production System that Tokyo Chuzosho developed fully incorporates the system's basic principles while tailoring the synchronization, equalization, and limited inventory/production specifications principles of the system to closely fit the mold of the company's current production system. However, Rome wasn't built in a day, and neither was Tokyo Chuzosho's ESP Production System.

In fact, the process of building ESP at the Gunma Plant involved a number of repeated studies concerning how this foundry company and its Gunma Plant could make improvements to simplify and reduce their characteristics and restrictions while introducing a more logical and better organized production system. Much time and effort were spent in resolving the improvement issues addressed by these studies. The restrictions at Tokyo Chuzosho and the Gunma Plant are described briefly below.

First, the casting processes posed various restrictions that affected production scheduling at the Gunma Plant. These complicated restrictions had made production scheduling a very difficult task. Consequently, the people responsible for production scheduling at the Gunma Plant were necessarily people who had a lot of expertise and experience in foundry technology and in the production resources and product items in the Gunma Plant.

Some of the items that fell under the category of production-scheduling restrictions at the Gunma Plant were upper limit values, foundry time segments, half-size assemblies, and molding sequences. Therefore, a production management logic was developed for this company's ESP Production System so that the company could prevent missing items and achieve highly efficient production while working within the bounds of the above-mentioned restrictions. The specific restrictions under each category of production scheduling restrictions are listed in Figure 8-5.

Category of restriction	Restriction
1. Upper limit restrictions	1. No. of flasks produced per day 2. Amount of molten metal poured per day 3. No. of production lots per day of products requiring cores 4. No. of flasks produced per day for special materials 5. Amount of molten metal poured per day for special materials
2. Molding time restrictions	1. Items that can be premelted 2. Items that can be finish melted 3. Allocation of special materials
3. Restrictions on half-size assemblies	1. Suitability of materials 2. Averaging of weight per flask 3. Equalization unit
4. Restrictions on molding sequences	1. Averaging of amounts of molten metal used 2. Balanced insertion of half size and full size

Figure 8-5. Production Scheduling Restrictions

Overview of Tokyo Chuzosho's ESP Production System and Production Scheduling Steps

Figure 8-6 shows a general flowchart of Tokyo Chuzosho's ESP system. In the figure, steps (A) to (F) indicate production scheduling steps that are performed every ten days, while steps (a) to (f) show more detailed production scheduling steps that are performed during each ten-day period. Steps (A) to (F) are particularly important because they determine what will be produced during the schedule's ten-day period.

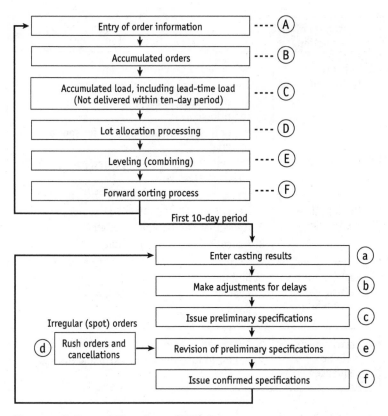

Figure 8-6. General Flowchart of ESP System

Production scheduling steps for ten-day schedules

Figure 8-7 illustrates the methods used at step (A) *Entry of order information* (Tokyo Chuzosho employees call this the *40-day planning step*). Under Tokyo Chuzosho's ESP Production System, order information for four ten-day periods is maintained. Order information for the first ten-day period is called *confirmed daily order information,* for the second, it is called *estimated daily order information,* and for the third and fourth periods, each is called *gross predicted ten-day order information.* All of this order information is entered into terminals by the sales staff and is transferred to external storage devices, after which the data is transferred every ten days to the host computer at the plant.

At Step (B), *Accumulated orders,* orders are tallied by the computer based on the transferred order information (see Figure 8-8).

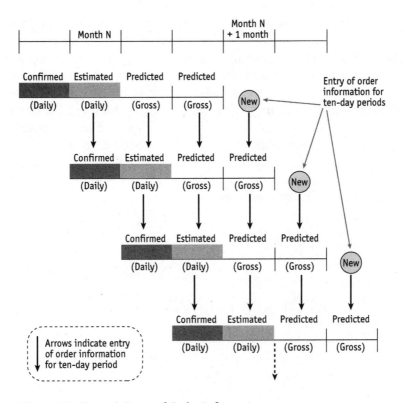

Figure 8-7. (Step A) Entry of Order Information

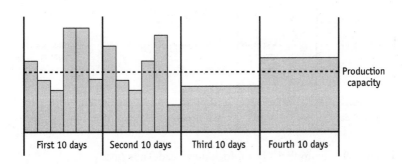

Figure 8-8. (Step B) Accumulated Orders

Next, the delivery deadline and delivery lead time are determined, and the load is fed forward so as to prevent missing items (based on a check of inventory levels). In other words, load is accumulated with the lead-time load included (see Figure 8-9). Only the lead-time amount is fed forward for any product item for which a ten-day limited inventory is being maintained (Product A in the figure), while the lead-time amount plus another ten-day amount is fed forward for spot-order items for which there is no limited inventory (Product B in the figure). This processing enables the factory to manufacture any product for which a ten-day load has been accumulated at any time during the ten-day schedule. In other words, within each ten-day period, the factory can maintain a production schedule with maximum production efficiency without having to worry about delivery deadlines.

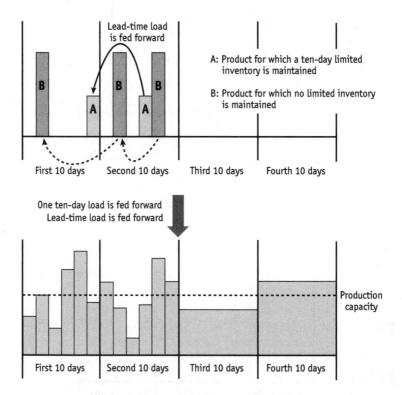

Figure 8-9. Accumulated Load, Including Lead-Time Load

Next, lot allocation processing is performed so that the same product items are not allocated to the same day (see Figure 8-10).

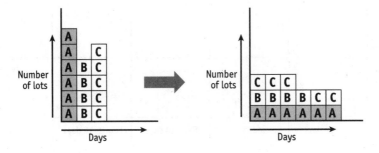

Figure 8-10. Lot Allocation Processing

In addition, Figure 8-11 shows how you do leveling (combining and leveling). This involves pairing products in order to meet the restrictions on combinations of materials as part of the lot allocation processing that occurs for half-size assembly production (combinations of two product items per flask).

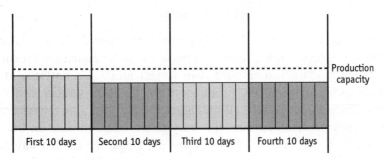

Figure 8-11. Leveling (Combining)

Finally, to maximize production efficiency, the products are sorted forward according to each day's maximum production capacity (see Figure 8-12). This is done by rearranging the production schedule for maximum production efficiency in each day. If there is any leftover capacity at the end of each ten-day period, production loads can be shifted forward from the next ten-day period to maximize production efficiency in each day. After shifting the production load, if a ten-day period ends up with one or more days with zero production scheduled, those days can either be used as scheduled downtime or production loads can be shifted forward from the next ten-day period.

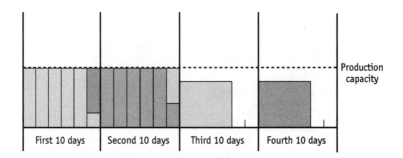

| First 10 days | Second 10 days | Third 10 days | Fourth 10 days |

Production capacity

Figure 8-12. Forward Sorting Process (Rearrangement)

The above types of processing are done for ten-day periods. Once you confirm the product items to be manufactured during the first ten-day period, you can work out a daily production schedule for that ten-day period.

Steps in planning a daily production schedule

The first step in planning a daily production schedule is to input the production results data. Next, you must manually adjust the schedule to accommodate last-minute rush orders and cancellations, after which you can issue the production specification vouchers. These steps, shown as steps (a) to (f) in Figure 8-6 on page 236, are repeated as necessary when planning the daily production schedule. In lieu of reading a description of each of these six steps, see the ten-day and daily production schedule planning steps outlined in Figure 8-13.

BENEFITS OF ESP AT TOKYO CHUZOSHO

Tokyo Chuzosho was able to realize many benefits from adopting the ESP Production System. The following list and the graphs in Figure 8-14 provide quantitative indicators of the benefits gained from ESP, as well as Tokyo Chuzosho's future goals. The *quantitative benefits* included:

- 23-point rise in delivery deadline achievement rate, raised to approximately 100 percent.
- 50-point rise in production schedule achievement rate, raised to approximately 100 percent.
- 83-point reduction in inventory volume, reduced to less than one-fifth of the benchmark value.

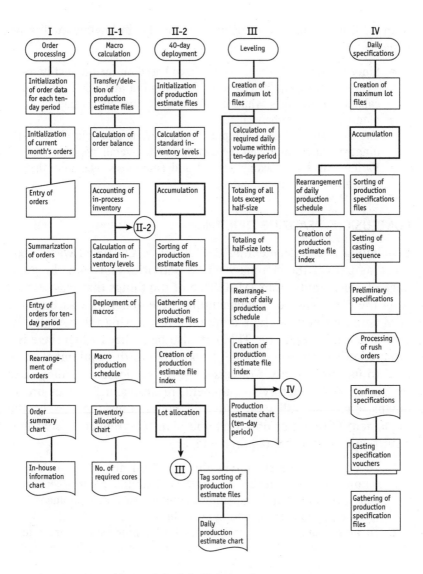

Figure 8-13. Production Schedule Planning Steps

- Reduction of lot size (increased number of changeovers), 3.6-fold increase in number of changeovers.
- Doubling of production efficiency in casting line.
- Reduction of 8 to 9 days per month in working days per unit of production volume, resulting in a major reduction in energy costs

In addition, Tokyo Chuzosho realized the following qualitative benefits.

- Since production units were equalized based on the product items (types), there were fewer causes for variation and this facilitated calculation of costs while having a major beneficial impact on plan management as well.
- The company's various computer systems were more closely coordinated, which improved the quality of the overall system.
- Various tasks were computerized and employees overcame their resistance to using computers.

CONCLUSION—REALIZING TRUE TPM

Developing and implementing ESP at Tokyo Chuzosho was undertaken as part of the company's TPM activities and with the guidance of consultants from JMAC. One of the things that was learned from the experience of implementing these activities was that true TPM activities must include more than just raising the overall efficiency of equipment. True TPM can also be realized when there is a production system (such as ESP) that enables the equipment to be used to its full capacity. Conversely, if the six losses are allowed to remain among a factory's equipment, resulting in frequent breakdowns, etc., you will gain little, no matter how thoroughly you implement ESP and or how well you plan the production schedule. Your two fundamental requirements are:

1. To eliminate the six losses through TPM activities such as autonomous maintenance activities, individual improvement activities, and skills acquisition activities, so that equipment is returned to its ideal operating condition.
2. To make improvements in manufacturing processes in order to realize a production system that is more efficient and builds higher quality into products.

Only when a company lays this foundation can the gears of the ESP Production System as a production management system operate like clockwork to create a truly efficient and profitable manufacturing organization.

After introducing ESP at Tokyo Chuzosho the customers are pleased, the company is profitable, and productivity has doubled! The authors encourage everyone at Tokyo Chuzosho to continue working together to achieve and maintain this type of organization through ongoing TPM activities that include the further improvement of their ESP Production System.

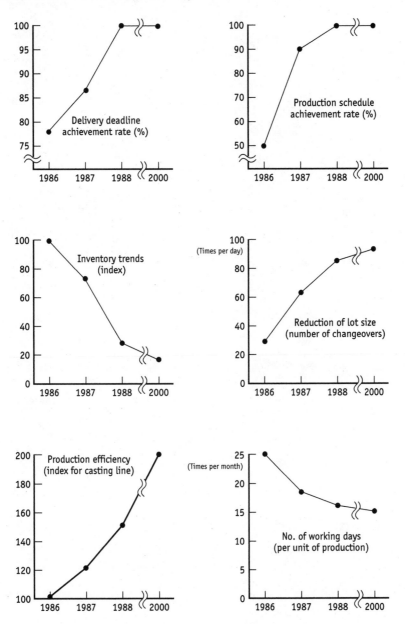

Figure 8-14. Benefits from Introduction and Deployment of ESP

Case Study 4: Plant A at Company N

Zero Defects Is Our Goal! and

No More Sloppy Inventory Management!

This case study concerns Company N's Plant A where production management concepts underwent a radical change under the ESP Production System. Company N was founded in 1956 as a manufacturer of sophisticated, high-precision measuring instruments. It got right to work contributing to labor-saving advances, with its air micrometers capable of 0.1 (micron) scale measurements, as well as various automated control devices. Later, Company N developed new technologies and conducted system engineering research in response to market needs, so that today the company is pushing back the frontiers in automation, labor reduction, and rationalization through its activities in five principal areas: fluid couplings, machine tools, linear compressors, electronic devices, and industrial robots.

OVERVIEW OF PLANT A AND THE CHARACTERISTICS OF ITEMS TARGETED BY ESP

Plant A's chief products are fluid couplings and linear compressors. Company N wanted to introduce the ESP Production System throughout Plant A. Consequently, this system was introduced and deployed for two main product categories: 1) fluid couplings and 2) linear compressors. Their key concerns with regard to customer service were to provide customer-satisfying quality and cost while ensuring on-time deliveries.

Production Processes at Plant A

Figure 9-1 shows an outline of the production processes at Plant A. Its two chief characteristics (constraints) are:

1. About 70 percent of machining processes are outsourced to supplier companies
2. Due to governmental policy, surface treatment processes are not nearby and, in fact, are not allowed to be included in the same factory.

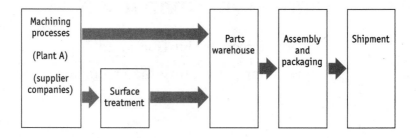

Figure 9-1. Outline of Production Processes at Plant A

Characteristics of Fluid Couplings

The fluid couplings targeted by ESP included some 2,000 product items as standard products. (This does not include special made-to-order products, which were not targets for the ESP Production System since they were manufactured only when specifically ordered.) To remain competitive with its standard products, Plant A needed to ensure prompt delivery for each of its standard product item numbers. The sales system for fluid couplings was based on using representative sales offices and warehouses for market distribution.

Characteristics of Linear Compressors

Company N manufactured two types of linear compressors: its own brand and an OEM brand. The sales system for Company N's own brand of linear compressors also used representative sales offices and warehouses for market distribution. And, as with fluid couplings, Plant A needed to ensure prompt delivery for each of its standard line of linear compressor products in order to stay ahead of the competition.

Orders for Company N's OEM-brand products were generally processed only once a month, but orders varied greatly in terms of the ordered product item numbers and quantities. Also, the company had to deal with frequent revisions of orders at the time of delivery, which meant that the production and delivery schedule had to be changed almost every day, despite the company's efforts to schedule production based on regularly timed orders. All linear compressor products were targets for the ESP Production System.

BACKGROUND AND PROBLEMS PRIOR TO INTRODUCING ESP

Given the harshly competitive market, Company N and Plant A were challenged to develop new products while boosting competitive factors such as cost and lead time to delivery. As is the case with many R&D-intensive manufacturers, Company N was turning out an abundance of new product items in response to new trends in orders and market needs, and consequently there was a huge assortment of product item numbers to be managed. However, even though they worked hard to develop and market new products, their management of existing products was relatively lax, which created all sorts of problems when it came to handling received orders, ensuring prompt deliveries, and controlling costs.

Eventually, Company N's managers became acutely aware that Plant A's products could no longer meet the QCD (Quality, Cost, and Delivery) requirements for remaining competitive in a diversifying marketplace. Accordingly, they began to seek ways to improve their companywide management system. In particular, they recognized that they would not be able to overcome their company's (and Plant A's) production management problems, which included chronic management-related loss, as long as Plant A outsourced 75 percent of its machining work to supplier companies, very few of which supplied products exclusively to Plant A.

The Problems at Plant A

Figure 9-2 outlines some of the problems faced by Plant A before introducing ESP and the goals established once it had been introduced.

Figure 9-3 lists factors related to the problems that Plant A was having before introducing ESP. These factors were considered from the perspective of the systems (information systems, functional systems, and physical systems) that had helped drive production.

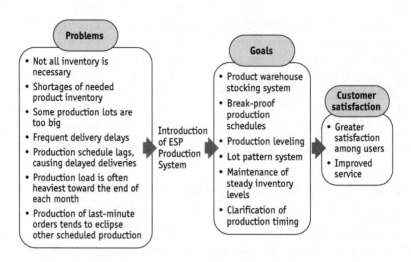

Figure 9-2. Problems Before Introducing ESP and Goals
Established Afterward

PROMOTIONAL ORGANIZATION FOR INTRODUCING ESP

The following three key points were formulated as guidelines for
establishing the promotional organization that supported the introduction of the ESP Production System at Plant A.

1. *Top managers will act as project leaders whose work will include
   making improvements.* Many managers strongly believed that
   their only job was to manage day-to-day production operations.
   It was therefore made clear to all managers that they were being
   made project leaders whose work would include introducing the
   ESP Production System. Also, as project leaders, these managers
   would help the project along in various ways, including keeping
   track of members' attendance at project activities and their hours
   spent on these activities.

2. *Project members will be assigned to improvement-promoting subcommittees involved in tasks other than their own work tasks.* For
   example, a subcommittee member whose daily work relates to
   fluid couplings may be assigned to a subcommittee whose focus
   is on linear compressors. Or, a subcommittee member whose daily
   work relates to machining processes may be assigned to a subcommittee whose focus is on improving assembly and packaging
   processes. This arrangement serves two main purposes: one is to
   exclude members who are professionals. The second purpose of
   this arrangement is to prevent people who are already experts
   in the target processes from downplaying the importance of

---

1) Production management: Information system-related problems and factors

1. Sales plans used target values for sales and manufacturing that were derived from past results for various product categories, and these types of sales plans were therefore not geared toward actual sales trends. Consequently, they were a factor behind imbalanced inventory levels.

2. Poor allocation control of warehouse inventory (products and parts) made it unclear which items could or could not be allocated.

3. Production load and capacity were not fully coordinated, which resulted in production schedules that exceeded production capacity, so that not all scheduled products could be manufactured.

   In particular, Plant A was unable to reliably estimate the load and capacity of the supplier companies that accounted for more than 70 percent of machining processes.

4. The Production Division was given only the main production schedule and production lots were indicated only in monthly amounts in the production specifications.

   Consequently, given the frequent occurrence of last-minute additions, deletions, and rearrangements in the production schedule, the main schedule and work instructions provided to the manufacturing lines were vague at best.

2) Production management: Functional (administrative) system-related problems and factors

1. Coordination was weak among sales planning, production scheduling, work instructions, and manufacturing functions. Often, products that were manufactured during overtime were found piled up in front of the plant and plant employees were not well informed about scheduling priorities.

2. Planning and purchasing functions were handled based on the corresponding processes or supplier companies, and overall process management functions were weak.

3. Various tasks were handled by "veteran" employees who basically performed only day-to-day troubleshooting.

   This resulted in a vicious cycle of "production change → production specifications (orders) that ignore capacity → delayed deliveries → imbalanced inventory levels."

3) Physical (production) system-related problems and factors

1. Lead times for purchased parts were too long. This was especially the case for parts that required surface treatment.

2. Categories of items to be manufactured at various processes (including in-house and at supplier companies) were vague, making it hard to keep track of their production routes.

3. Parts packages and quantities were not standardized and were therefore inconsistent. Some items would end up staying in parts warehouses for long periods.

4. The plant lacked the production capacity to meet occasional needs for higher production output.

Figure 9-3. Systems-based Perspective on Factors behind Problems Prior to Introduction of ESP Production System

studying current conditions and other preliminary steps toward making improvements. Experts can easily overlook or misunderstand the true causes of various problems. By contrast, subcommittee members who know little about the target processes tend to follow problem-solving procedures more strictly and are more attentive and open-minded when studying current conditions. Thus, they are more likely to carefully implement every step toward making improvements and to identify the true causes of problems.

3. *Subcommittees will be established for specific products, product types, and processes.* Establishing several subcommittees for the same purpose or goal enables the various subcommittees to prod each other toward better understanding of problems and implementation of better improvements. This arrangement also puts the principle of competition into effect, which helps the subcommittees keep to their project activities schedules and reduces variation in improvement steps and procedures.

POLICIES REGARDING INTRODUCING AND DEPLOYING ESP

Plant A's top managers devised the following policies concerning the introduction and deployment of the ESP Production System at their plant. These policies were used as guidelines when fleshing out the details of the project. Having formulated these policies at the beginning of these activities, the top managers were then able to clarify the project's direction. The policies also helped project members make clear and confident choices on their course of action when confronted with obstacles or when having to choose among several proposals.

1. Ensure prompt delivery service and do not pass demand fluctuations directly to the Production Division.
   - Use product inventory (including distribution inventory) to create specific separation effects between demand and production.
   - Due to these separation effects, the Production Division will be able to level production and achieve a steadily high operation rate. The ESP Production System provides techniques for doing this.
   - Establish and administer rules and procedures for drafting sales plans, production schedules, and inventory schedules.
   - Minimize inventory levels. Basically, the plant should be run without any parts inventory.
2. Implement thoroughly stratified management.
   - Stratification of quantities.
   - Stratification (grouping) of similar products.

3. Do not deploy new methods for all products or item numbers at the same time. Instead, select a particular model and succeed with that one before introducing other methods.
   - Selecting particular models will also help prevent problems and confusion that might otherwise occur later during full-scale deployment.
4. Make improvements in physical systems to enable execution of ESP production management.
   - Regulate the flow of objects.
   - Make improvements to establish ideal conditions rather than conditions that merely comply with standards.
   - Seeking ideal conditions will enable improvements and standards to be established more quickly and maintained more easily in the future.

STEPS IN DEPLOYING ESP AT COMPANY N'S PLANT A

There were four overall stages encompassing the steps for deploying ESP. In Stage 1 the company analyzed the physical systems (to clarify true constraints); in Stage 2 they established planning standards; in Stage 3 they created the basic master plan; and in Stage 4 they established methods (see Figure 9-4).

ESTABLISHING RESULT MEASUREMENT INDICATORS AND TARGET VALUES

The objective in making improvements in Plant A's ESP Production System was to achieve six things.

1. Ensure prompt deliveries of zero missing items.
2. Prevent overproduction.
3. Establish leveled production through coordinated scheduling.
4. Maximize production efficiency.
5. Flexibly respond to production changes.
6. Establish production management logic, a departure from decision making based on an inner circle of managers.

In order to quantitatively measure whether or not these were being achieved, they needed to establish results indicators and target values (see Figure 9.5).

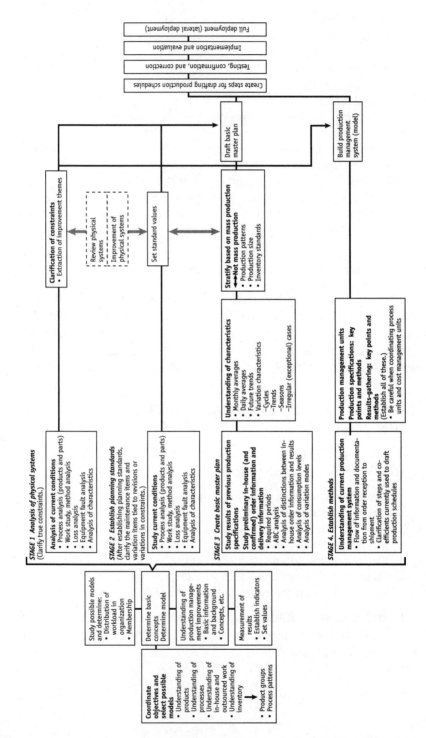

Figure 9-4. Stages to Deploying ESP

Results indicator		Formula	Target value
1. Product sufficiency rate		Target month's total shipments / target month's total received orders x 100	100%
2. Production schedule adherence rate	1) Completion rate for production items	Scheduled items completed in target month / total scheduled items in target month x 100	100%
	2) Synchronization rate	Scheduled items completed by specified date / total scheduled items in target month x 100	100%
	3) No. of days delayed	(Specified date for completion of production schedule − Date when production is completed) / Items not completed by specified date	0 days
3. Inventory trends

* Indicators marked with asterisks are used to understand both:
• Volume
• Value | 1) Product inventory | Target month's product inventory / product inventory used as reference x 100 | ▲30% |
| | 2) In-process inventory | Target month's in-process inventory / in-process inventory used as reference x 100

*These are studied separately for machining, surface treatment, and assembly and packaging processes. | |
	3) Parts inventory	Target month's parts inventory / parts inventory used as reference x 100	
	4) Materials inventory	Target month's materials inventory / materials inventory used as reference x 100	
4. Production efficiency	1) Overall equipment efficiency	Machine time x No. of nondefective goods / operation time x 100	85% or above
	2) Operator productivity	Standard time x No. of nondefective goods / operation time x 100	100%
5. Changeover frequency trends (conditions when shifting to smaller lots)		No. of changeovers in target month / number of changeovers used as function x 100

*These are studied separately for machining and assembly and packaging processes. | 300% |

Figure 9-5. Results Measurement Indicators and Target Values

ESP PRODUCTION SYSTEM FOR FLUID COUPLINGS

First, there was the selection of model products. This entailed stratification of volume and of product types. As a result, the products were divided into two main types, which are discussed below.

1. *Stratification of volume.* ABC analyses were performed based on a survey of sales results, and product types were ranked according to their production volumes.
2. *Stratification of product types.* The products manufactured at Plant A were first stratified based on three categories: model, size, and major parts and materials, after which they were stratified again based on their minor parts and materials.

Products were then grouped according to their first, or parent, layer of stratification and their second, child, layer of stratification. In addition, they identified parts that were common among multiple product groups, studied the relations among these parts, and stratified products based on those relations. The result of all of this stratification was that products were divided into two main types.

1. *Type I: Low-variety, large-lot production models.* These are Plant A's main products, which include several dozen item numbers among the finished products. All of these products use standardized parts and are assembled by industrial robots.
2. *Type II: Wide-variety, small-lot production models.* These products are selected from among the remaining item numbers (of which there are more than one thousand).

For most of these products, details such as where, when, and how each product was assembled remained unclear. Next, they hypothesized a line of assembly processes and selected (as models) products that they could manufacture in a mixed production system using this assembly line. All together they selected 200 item numbers among eight model products.

IMPROVEMENT MEASURES TO ESTABLISH ESP

First we will discuss the improvement measures for both Types I and II, then the improvement measures for Type I only, and then Type II only.

Nine Improvement Measures for Type I and II Production Models

1. Large lots produced on a monthly basis were divided into smaller lots with shorter time periods.
2. The scheduling cycle was shortened from once per month to once every three days and production was managed using a three-day cycle, 12-week rolling schedule. The Production Division provided production specifications for the following three ways, (items 3 to 5) but did not make any subsequent changes in those specifications.
3. Lot sizes (equalized units) were determined per item number and these equalized units were used when scheduling production at machining processes and assembly processes.
4. Safe product inventory levels were established to ensure prompt delivery of products
5. Upper limits and lower limits of safe inventory levels for products were set and signaling methods to prevent shortages or excess inventory were devised.
6. Machining processes were reorganized, the rectification of production and the coordination of production load and capacity were clarified, and factory-floor improvements were made to boost production capacity.

- In the case of Plant A, since over 70 percent of the machining processes were being outsourced to supplier companies, each company that performed machining was viewed as a specialized line that was allocated to particular product groups according to factors such as their production capacity, the scope of their transactions with Plant A.
- The methods used for these tasks included changing certain product orders, moving production of certain items in-house, and establishing ties with new supplier companies.
- These supplier companies faced their own problems, but were in some ways helped at first by a favorable business climate. As a result, they were able to coordinate their respective production loads and capacities and were positive about the new outsourcing system once it had become established.

7. Purchasing lead time was shortened.
   - It enabled parts to be transferred directly from machining processes to surface treatment processes.
   - It doubled the frequency of the conveyance cycle, from once per week to twice per week.
8. The packaging of delivered parts was modified to smaller parts containers, and more consistent quantities of parts were established.
   - They switched from cardboard containers to plastic pass-box type containers.
   - They equalized quantities of parts in containers by requiring all parts quantities to be an integral multiple of the parts required per product lot.
9. The order-point method and double-bin method were used to manage common-value parts or other common parts at assembly processes.

Three Improvement Measures for Type I Production Models

These measures helped build a system that prevents inventory shortages or surpluses through rhythmical production of low-variety, large-lot models arranged to maximize economies of scale.

1. A production sequence was established for each item number and then production patterns were established in which item numbers were coordinated with equalized units. Once the production orders are confirmed, these production patterns are used to draft a production schedule. If the production orders are subsequently changed, these production patterns can be rearranged accordingly, instead of rearranging the individual production orders.

2. Production managers can respond to demand changes (changes in production orders) by rearranging the equalized units in the production schedule.

3. The production patterns are reviewed every three months.

Three Improvement Measures for Type II Production Models

Type II (wide-variety, small-lot production) models were established and arranged. Each product group was further broken down into three groups: 1) items for scheduled production, 2) items to replenish inventory levels, and 3) items for spot production (production of last-minute orders). Next, equalized units were established for each item number. Once these were elucidated, managers were able to engage in frame planning of production capacity in advance.

1. The group called *items for scheduled production* were assigned production frames based on major item numbers. The parts that belonged to each of these major numbers were identified by converting them from the production frames of major item numbers using equalized units each time a production order was issued.

2, The group called *items to replenish inventory levels* was used to schedule assembly of equalized units of parts taken from the parts inventory whenever an order to replenish the product inventory was issued. In view of the lead time for purchasing of parts, whenever an order to replenish the product inventory was issued, replenishment of parts in equalized machined quantities was ordered from in-house and outsourced machining processes.

3. In addition, production managers helped level production by scheduling production of seasonal products ahead of their peak seasonal demand periods.

Making Improvements for Type I Production Models

The first order of business was to establish equalized units for each item number (see Figure 9-6). Then they studied sales results, created a production cycle, and equalized units, which are discussed next. Then they needed to create the production patterns, set limited inventory levels, and, finally, attend to tasks related to production scheduling and production specifications.

1. *Study of sales results.* The past year's day-to-day sales results were studied to get a better understanding of sales conditions. This

information was used as reference material for establishing equalized units and limited inventory levels, as described below.

2. *Create production cycle (proposal).* After performing ABC analyses and variation studies of sales, proposals were drafted for daily production, weekly production, three-times-per-month production, twice-per-month-production, monthly production, and order production.

3. *Create equalized units (proposal).* After performing ABC analyses of sales results and calculating required volumes, proposals were drafted for equalized units. At that time, one key point was the proposal for smaller lot sizes, based on the monthly production and weekly production proposals. The project members also wanted to be sure to take packaging and equalized production into consideration.

   • The equalized units were determined based on the sales results (required volumes), production cycles, assembly methods, production capacity, and changeover times. However, some research remains to be done concerning logical setup methods.

   • In this case study, equalized units were set as daily units, but for some types of production it may be possible to established hourly units (to create even smaller lot sizes).

Product item number	Sales results					Equalized units for specific production cycles				
	Maximum volume	Minimum volume	Average	Rank	Variation	Daily	Weekly	3 times/ month	2 times/ month	Monthly
A										
B										
C										

Figure 9-6. Example of Equalized Units Set for Individual Item Numbers

Creating production patterns

Figure 9-7 shows an example of a production pattern. This entailed setting a production sequence and creating production patterns using equalized units, production cycles, and production sequences. These are discussed below.

1. *Setting production sequence.* Project members studied quality requirements (such as preventing assembly errors or use of incorrect parts) and changeover requirements, then they organized this information and set the production sequence. As for

changeovers, they studied changeover improvements that would combine item numbers that were seen at that time as requiring longer changeover times. After setting the new production sequence, they soon implemented changeover improvements to combine item numbers that had required longer changeover times. Later, improvements related to quality requirements and changeovers would be taken into consideration during periodic reviews (every three months) of the production patterns. (*Note:* see the third improvement for Type I, described on page 256.)

2. *Creating production patterns using equalized units, production cycles, and production sequences.* Project members created production patterns using the equalized units for specific item numbers and production cycles. At that time, order (spot) production items were not included in production patterns. Standard item numbers were checked for compatibility with other item numbers and groupings were devised and then organized as information.

Product item number	Year N, Month N																
	1st (Mon)	2nd (Tues)	3rd (Wed)	4th (Thurs)	5th (Fri)	6th (Sat)	7th (Sun)	8th (Mon)	9th (Tues)	10th (Wed)	11th (Thurs)	12th (Fri)	13th (Sat)	14th (Sun)	15th (Mon)	16th (Tues)	17th (Wed)
A	←→							←→			←→						
B			↔							↔					↔		
C				↔												↔	
D				↔													↔

Figure 9-7. Example of Production Pattern

Setting limited inventory levels

The next undertaking to make improvements for Type I was to set limited inventory levels. This entailed: 1) using production patterns to simulate inventory input, output, and storage; 2) setting a lower limit for limited inventory; and 3) setting an upper limit for limited inventory. These are discussed below.

1. *Using production patterns to simulate inventory input, output, and storage (see Figure 9-8).* To get a realistic grasp of demand fluctuations, project members performed simulations of the inventory input, output, and storage of specific item numbers. They used inventory amounts indicated in production patterns to perform the inventory input simulations and they used sales results to perform the inventory output simulations. The results of

these simulations were used to review the production cycles, equalized units, and production patterns. Also, the largest negative value shown for inventory storage in the simulations was set as the *amount absorbed by demand fluctuation.*

For the amount absorbed by demand fluctuation, they did safety coefficient calculations using the out-of-stock-rate as a quantitative reference, but in the end they obtained smaller values from the preliminary simulations described above. Using simulations also made their results more impressive to people in other parts of the company. Project members believed that this was due to emphasizing the requirement that *prompt delivery equal zero missing items* as part of their calculations using safety coefficients. They decided they would need to revise their calculations henceforth.

Product item number A	Year N, Month N																
	1st (Mon)	2nd (Tues)	3rd (Wed)	4th (Thurs)	5th (Fri)	6th (Sat)	7th (Sun)	8th (Mon)	9th (Tues)	10th (Wed)	11th (Thurs)	12th (Fri)	13th (Sat)	14th (Sun)	15th (Mon)	16th (Tues)	17th (Wed)
Input	10	10	0	0	0	–	–	10	10	0	10	10	–	–	0	0	0
Output	3	7	6	0	5	–	–	17	8	0	4	7	–	–	0	9	3
Storage	7	10	4	4	▲1	–	–	▲8	▲6	▲6	0	3	–	–	3	▲6	▲9

Figure 9-8. Example of "Inventory Input, Inventory Output, and Inventory Storage" Simulations based on Production Patterns

2. *Setting a lower limit for limited inventory.* The formula used to set the lower limit for limited inventory is: **scheduling cycle (in this case, a three-day cycle) × average demand volume + amount absorbed by demand fluctuation.** When the actual inventory amount goes below the lower limit, a production order signal is issued to request production in equalized units. This helps to maintain the prompt-delivery system.

3. *Setting an upper limit for limited inventory.* The formula used to set the upper limit for limited inventory is: **lower limit amount + equalized units.** When the actual inventory amount goes above the upper limit, a production stop order signal is issued. This helps to prevent surplus inventory.

Tasks related to production scheduling and production specifications

1. The fixed period for production scheduling was set at three days. This means that production specifications for the next three days

would be issued each day. These production specifications are part of the confirmed schedule issued to each manufacturing facility, which means that once they are issued they cannot be changed. Thus, every day a nonrevisable, confirmed three-day production schedule is established. This company strictly enforced the rule that changes in the production schedule could be made only by production managers and only before these production specifications were issued to manufacturing facilities.

2. Any improvements made in a manufacturing plant must be based on the production scheduling standards and the standards for executing production management tasks. As a result, relations have been established among the functions performed by production management tasks. In Figure 9-9 the parts enclosed in shaded boxes are tasks that have been standardized and programmed to enable processing by computers.

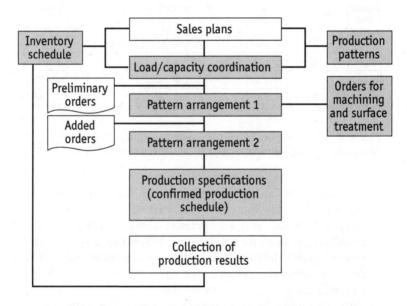

Figure 9-9. Basic Sequence of Production Management Tasks

Making Improvements for Type II Production Models

As with improvements for Type I, there were many steps involved. The project members needed to establish production triggers for specific item numbers, set up product group configuration and coordination of production triggers for product groups, and provide the administration of the scheduled production group, the inventory replenishment group, and the order (spot) production group.

Establishing production triggers for specific item numbers

Using the sales results, project members stratified all Type II items into three categories: 1) items for scheduled production, 2) items to replenish inventory levels, and 3) items for spot production. After studying daily sales results from the past year to get a grasp of sales conditions, they used this information as a reference for establishing equalized units and limited inventory levels, as described below. They then set a provisional production cycle and provisional equalized unit for specific item numbers, based on the sales results.

Product group configuration and coordination of production triggers for product groups

All Type II item numbers comprise the first category (model, size, and major parts and materials) as a parent group. Next, all item numbers belonging to the parent group comprise the second category (minor parts and materials) as a child group. The item number in each product group (child group) that has the largest production volume is used as the representative item number. The production trigger for each representative item number is used as the production trigger for the corresponding product group (see Figure 9-10).

Product group (trigger for representative item number)	Production trigger for specific item number
Scheduled production group	Group of items for scheduled production
	Items to replenish inventory levels
	Items for spot production
Group of items to replenish inventory levels	Items to replenish inventory levels
	Items for spot production
Group of items for spot production	Items for spot production

Set for each child group

Figure 9-10. Product Groups, Corresponding Production Triggers, and Production Triggers for Included Product Item Numbers

Administration of scheduled production group

1. *Setting production triggers for specific item numbers.* First, project members used the shortest production cycle for the product item numbers in the product groups to recalculate the (provisional) equalized units for specific item numbers, and then they added up the results for each product group. Next, they used this information as a basis for planning machining capacity and the capacity of parts containers and product containers, and for setting equalized units for each product group. These equalized units for each product group were in turn set as the machining lots (i.e., ordering units).

2. *Administration of scheduled production group.* For the scheduled production group, the equalized units for each product group were used as the volumes for frame planning. As production of scheduled production items continued, production request signals for items to replenish inventory levels and/or items for spot production within that group were used to allocate orders within each frame as part of the process for drafting production schedules.

   If the target group didn't include any orders for items to replenish inventory levels or items for spot production, only scheduled production items would be manufactured. This method of administration helped reduce parts-inventory levels while broadening the scope of load adjustments for the future. Figure 9-11 shows an example of an inventory input/output form used for administration of the scheduled production group.

Date	1st	2nd	3rd	4th	5th
Parts No. of parts input No. of parts output No. of parts in inventory					
Parent item numbers Basic production volume No. of production specifications Sales volume Amount in product inventory					
Child item numbers No. of orders No. of production specifications Sales volume Amount in product inventory					

Figure 9-11. Example of Inventory Input/output Form Used for Administration of Scheduled Production Group

3. *Setting limited inventory levels.* For the scheduled production group, the production cycle and the equalized units for specific product items were used to perform inventory input/output simulations to get a better understanding of the amounts of demand fluctuation. The following formula was used to calculate the lower limit of inventory for specific item numbers. You always calculate the upper inventory limit as an integral multiple of the equalized units. The formula for calculating lower inventory limit of items for scheduled production and items to replenish inventory levels is: **scheduling cycle (in this case, a three-day cycle) × average demand volume + amount absorbed by demand fluctuation.** Items for spot production is: **no inventory**

Administration of inventory replenishment group

1. *Administration of inventory replenishment group.* For this group, in order to maintain the necessary inventory of parts, a lower limit (reorder point) was set for product inventory so that the necessary parts would be produced (assembled).

2. *Product inventory standards and assembly amount specifications.* This company's production planning cycle for assembly was set as a three-day cycle. Adding the amount absorbed by demand fluctuation to the average demand volume for three days set the limit (reorder point). In other words, the formula used for lower inventory limit was: **lower inventory limit = (average demand volume for three days) + {(maximum sales volume per day – average demand volume) × 3}.**

   In addition, they set the upper inventory limit by adding the assembly volume (in equalized units) to the lower inventory limit value. At each production planning cycle (every three days) the effective inventory volume was checked and the amounts of any products that were below the lower inventory limit were calculated by subtracting the effective inventory from the upper inventory limit. Then this amount was rounded off as equalized units for assembly, after which assembly work instructions were issued to start production.

3. *Setting equalized units for assembly (for specific item numbers).* The part package counts (units per container) are equalized units for product assembly. This virtually eliminates the need for management of extraneous parts (remaining parts that are not enough to fill a container). The key point here is to establish small lots for delivery of parts. When parts are delivered in minimized lot sizes,

it adds flexibility to assembly processes while helping to minimize product inventory levels.

4. *Parts inventory standards and machining volume standards.* Amounts of parts inventory are set for each product group (those having common major parts). The purchasing periods and sales results for major parts are used as a basis for the calculation: **purchasing period** × **average demand volume (total for product group) + amount absorbed by demand fluctuation.** This value is then rounded off according to the package count for parts deliveries to obtain the lower inventory limit (reorder point).

The machining volume calculation is: **the purchasing period (rounded off)** × **average demand volume (total for product group).** The calculation is performed according to the package count for parts deliveries, and the amount ordered each time is then equalized.

Administration of order (spot) production group

The order production group consists of products that have a clear production lead time, which is reported to the customer when the order is received, and these products are delivered to customers within the reported lead time. One incidental effect of this approach is that products that were previously considered special order items are now included with standard items.

Quantitative Benefits of ESP for Fluid Couplings

To sum up the improvement process for levels I and II, Figure 9-12 shows the company's quantitative benefits of using ESP for fluid couplings.

ESP PRODUCTION SYSTEM FOR LINEAR COMPRESSORS—CONDITIONS BEFORE ITS INTRODUCTION

With the previous production scheduling system, the production schedule was frequently augmented or otherwise revised, which caused variation in product models and production lot sizes and required a considerable amount of labor hours for monthly production scheduling. For this company, the objective in introducing the ESP Production System was to liberate themselves from their current production planning system and put into place a system that would make production much more efficient.

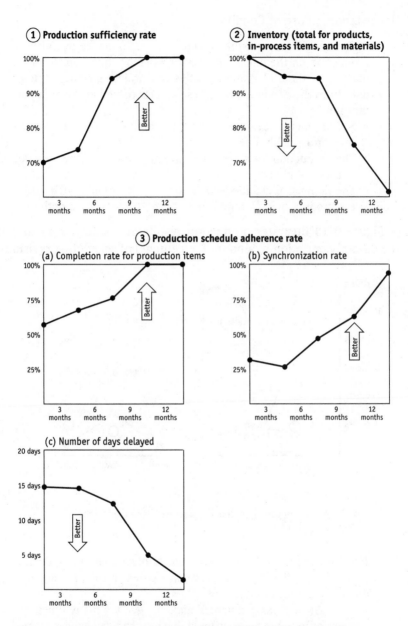

Figure 9-12. Quantitative Benefits of Using ESP for Fluid Couplings

During the first year, the main objective was to introduce and firmly establish the ESP Production System. A target of boosting production efficiency by 10 percent was also set for this year.

Understanding Current Conditions

1. *Overview of linear compressor production.* To get an overview of production on the linear compressor assembly line prior to intro-duction of the ESP Production System, project members studied data from the previous year. These studies turned up three impor-tant points.
   - Six product models were being assembled on the linear com-pressor assembly line.
   - The production lot sizes varied greatly, from 10 items to 2,000 items per lot.
   - The number of changeovers (retooling for assembly of a differ-ent product model) varied from 11 to 26 times per month.

Figure 9-13 llustrates production results and Figure 9-14 graphs the monthly variation in amount (percentage) of scheduling revisions.

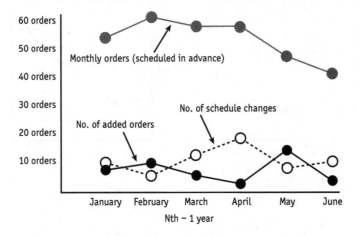

Figure 9-13. Production Results (Data from Year Prior to the Introduction of ESP)

2. *Problems in previous production scheduling system.* The following points summarize the problems that existed in the previous pro-duction scheduling system.
   - The schedule was augmented and changed too many times. The types of loss that were suffered due to schedule changes created the need for the following types of tasks and operations.
     - Rewriting the schedule
     - Rechecking required parts
     - Issuing new orders for related parts
     - Replacing subassembly schedule

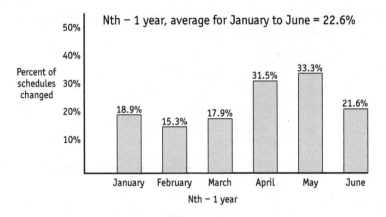

Figure 9-14. Monthly Production Schedule Changes

- Psychological impact on line operators
- Other
• Variation exists among scheduled lots.
   - Variation exists in labor hours (work load) for monthly scheduling.
   - Little attention is paid to labor hours (capacity).
   - Scheduled labor hours are not stable.
   - Too many changeovers. Changeover (retooling for assembly of a different product model) was being performed three times a day.
   - Scheduling was customer-centered. Schedules were planned strictly for the required volumes.
   - Since inventory was considered a bad thing, the production schedule had to be interrupted whenever spot orders were received.
   - Production efficiency was not being emphasized.
   - Production schedules were being planned at the company's head office, away from the manufacturing plant.

For the inventory replenishment group, a parts inventory was maintained and production (assembly) was ordered whenever the product inventory level went below the lower limit (reorder point).

Major Improvement Items

1. *Switched from monthly schedules to 10-day schedules.* Previously, this company's head office had scheduled production on a monthly basis. The improvement switched the location of production

scheduling to the manufacturing plant and switched the timing to every 10 days. (See Figure 9-15.)

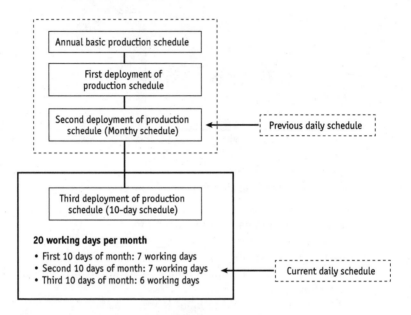

Figure 9-15. Deployment of Production Scheduling

2. *Introduction of production patterns as part of ESP Production System.* Previously, this company had to schedule production to meet the volume and delivery requirements of combined customer orders for six different product models. This improvement established equalized units for each product model, based on studies of previous sales results. It also established production input sequences that facilitated changeovers (retooling for model changes) and determined production patterns (such as shown in Figure 9-16) to maximize production efficiency. These production patterns are used to administer the ESP Production System as a production system based on highly efficient production scheduling. Some key points of this improvement are:
   - Similar models are produced together.
   - Production lot sizes are scheduled based on equalized units.
   - Production patterns are used.
   - Setting production requires an adjustment date. (This made possible extra production capacity resulting from the ESP Production System.)

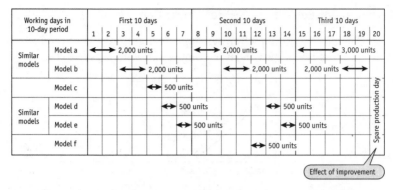

Figure 9-16. Example of Production Pattern for Linear Compressors

3. *Changed the timing for determining production schedules (see Figure 9-17).* Before introducing the ESP Production System, each monthly production schedule was set up during the previous month. Since then, each production schedule is for a 10-day period and is set up (as an arrangement of production patterns) during the previous 10-day period. Once a 10-day schedule has been established and the corresponding specifications (instructions) have been sent to the manufacturing plants, the production schedule must not be changed. This means that only confirmed production schedules are used at manufacturing plants, and production managers can trust that these confirmed schedules will remain unchanged.

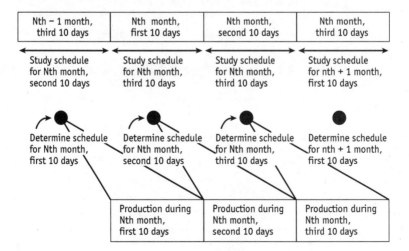

Figure 9-17. Timing of Determining Production Schedules

Improvement Benefits of ESP for Linear Compressors

1. Percentage of production schedule changes outside of production patterns (see Figure 9-18 in comparison with Figure 9-14).

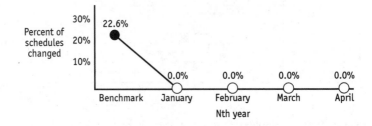

Figure 9-18. Improvement in Monthly Schedule Changes After ESP

2. Number of changeovers per month (see Figure 9-19)

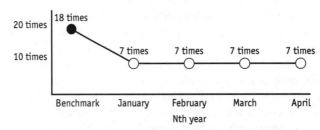

Note: *Since changeovers of similar models involve only changing of parts, they are not included in the number of changeovers indicated in the benchmark and in the results for the Nth year.*

Figure 9-19. Improvement in Number of Changeovers per Month After ESP

3. Comparison of previous system and ESP Production System (see Figure 9-20)

	Previous production system	ESP Production System	ESP's advantages
Production system	Order-based production system	Order-based production system + market production system	• Production schedule can be arranged farther in advance • Enables medium and long-term business planning
Production methodology	• "Shish kebab" style production using large lots • Irregular periods and undefined amounts	• Synchronized and equalized production using small lots • Regular period and stable amounts	• Enables use of patterns based on standard-size lots • Production is more leveled and stable, making for higher production efficiency • Production capacity can be more easily understood • Facilitates production scheduling
Production cycle	• Up-and-down, unstable cycles	• Leveled, stable cycles	
Lot size	• Unconfirmed	• Confirmed	

Figure 9-20. Advantages of the ESP System

CONCLUSION—OBSTACLES ENCOUNTERED WHILE IMPLEMENTING ESP

First, when introducing the ESP Production System, it is important to gain cooperation from various departments and to have a flexible Information Systems Department. At Company N, computerization was promoted to help the company permanently free itself from the old inner circle approach to production management. However, project members encountered obstacles in the form of lack of cooperation from various departments in the gathering of results data and in the use of information systems. This was especially true during the trial-and-error stage where basic data was being collected and new logic was being developed as part of information-related improvements. The company was forced to make several adjustments in their schedule for ESP Production System introduction because of these obstacles. In this case, the project members at Company N became painfully aware of the need for good cooperation among departments and flexibility on the part of the Information Systems Department as a backup organization.

Second, you must have strict adherence to the rules in order to maintain and improve the ESP Production System. Any new system has rules that must be followed if the system is to be maintained and expanded. Japanese employees have been noted for their spirit of cooperation and their willingness to put personal considerations aside for the good of the organization, and this characteristic will be needed in the future along with a willingness to be frank about problems and to search for their true causes. This is also something that Company N's project members came to appreciate.

Considering the many factors that go into a production management system, such as service characteristics, product characteristics, the company's history, process characteristics, and geographical conditions, it is impossible to say that any particular production system is the right solution for all companies.

To further improve their production system (such as the ESP Production System), employees at each company must have a powerful and clear determination to continually look for ways to make new improvements and reach new goals. Without such firm resolve on the part of employees, the production system will eventually become mismatched with the company's needs.

Also, no company can hope for much success if they merely copy the superficial aspects of successes achieved by other companies in improving their own production systems.

The year spent introducing the ESP Production System at Company N was one in which everyone involved learned the critical importance of establishing basic policies and approaches to goals resolutely and not giving up until the goals are achieved.

Glossary

Autonomation: The English translation of *Jidoka*. The idea came from an automatic weaving machine invented by Toyota Sakichi (1867–1930). He invented a machine with an automatic stoppage device that ceases the production if it detects problems in machine or defects in products. This was later developed as Jidoka (Autonomation), meaning attaching an automatic stoppage device into production machineries to prevent defects and losses. Autonomation enables one person to monitor multi-machineries; therefore it saves manpower, and it is one of the two pillars of Toyota Production System.

Buyer: In ESP, a buyer is a purchasing department or a person who buys and orders parts and materials. In this book, it indicates a purchasing department in a customer company, and a purchasing department in its own company.

Changeover loss: Loss caused by a production stoppage when changing a set-up of production line for different product types or item numbers. Changeover loss is a time loss between the time when production of the previous product is stopped and the time when a standard operation is resumed for the next product.

Clear planning standards: To clarify various conditions of a standard of production schedule. Those conditions include:

- Production conditions of overall production process:
  - Operation time
- Number of employees, attendance record
  Overall equipment, equipment breakdown rate
  Processing process standard
  Assembly process standard
  - Preparation and supply of parts and materials (supply by own company or by a supplier)
  - Order unit and delivery time of parts and materials (lead-time), and defective rate
- Production conditions of each product type or product item number
  - Workload standard
  - Planning standard
  - Parts and materials composition
  - Defective rate, yield rate
  - Changeover man-hour
  - Inventory standard

Controlled minimum inventory: The smallest possible inventory to conduct a production without any missing products. "Controlled" means to maintain a minimum inventory standard.

Customer: In this book, a customer means a company or factory that buys our products. Customer and delivery site have the same meaning.

Dead inventory: Inventory of products, parts, and materials that are unlikely to sell anymore.

Defect loss: It means a quantity loss of materials. There are two kinds of defect loss:

- Process Defect Loss
  - Quantity loss caused by defects and readjustment. Readjustment of products requires re-processing and it is considered as a loss of time and manpower.
- Start-up loss and yield (decline of yield) loss
  - Start-up loss is a loss in quantity occurred between the production start-up time and the time production is stabilized. Yield loss is a loss in quantity occurred when yield rate is below the standard yield.

In addition to the loss of quantity, defect loss also includes a time loss for production time used for defects and readjustment.

Delivery sites: A place where products are delivered (e.g. Customer's storage house, designated area in a production line, etc.)

Direct division: In this book, a direct division means a manufacturing department. Divisions other than direct division include Management division (production control, purchasing management, distribution management), Engineering division (design, production engineering, etc.), Research and Development division, Marketing, and other administrative divisions.

Direct operations: Direct operations are actual manufacturing operations, such as processing, assembly and equipment operation, etc). In this book, transportation and preparation of parts and materials and management of production line are categorized as semi-direct operations.

(Eliminate, Combine, Rearrange, Simplify): Four principles of Kaizen— Eliminate, Combine, Re-arrange, and Simplify.

End buyers: Final buyers. A department or person to whom a supplier delivers the final product.

Enterprise resource planning (ERP): ERP is to unify and control various management resources (people, product, money and information) of the company. Recently has been common to create "ERP system package," that includes production, supply/procurement, inventory, sales, finance, accounting, and human resources. The purpose of implementing the ERP system package is to grasp the actual company's activities, establish a system for worldwide competition, implement the latest operation know-how, and construct an information system quickly.

ESP highway items: Also called an ESCORT highway. In ESP, categorizing products (product type and product item number) is an important element. ESP highway items are products that have limited inventory. These items are made with limited inventory, and it guarantees a quick product delivery to customers.

In general, ESP highway items are products that are ordered continuously in a certain amount. In Japan, it is also called "Joryuhin (usual flow items)."

ESP information system: Information system that has ESP production planning functions (order management function, shipping management function, inventory control function, production planning standard, production planning simulation and logic planning function), ESP parts/materials ordering function, ESP production order function, ESP production result gathering function.

ESP production pattern: Also called an ESCORT production pattern. ESP production pattern is a preliminary production plan that helps the company clarify its production planning standards, as well as anticipate demand from buyers. ESP uses the ESP production pattern as a temporary production schedule. To be concrete, for each production line, the production sequence and volume per production run of each product variety and product item number have determined, and produced repeatedly in a certain cycle.

Equalization: Equalization means that "ordering and receiving the supplied goods," "production unit of each product item number," and "supplying parts and materials according to the production unit of each product item number" are done in the same volume. Equalization is an original conception of ESP, and it enables the achievement of the "Six Guarantees" and greatly reduces complicated volume/quantity management in the production process.

Equalized units (equalized purchasing units): Fixed unit volumes of order, delivery, transportation and production of supplied parts and materials and production of each item number.

Feed-forward production load: It means "production smoothing (heijunka)" in ESP. The general method is, first, to understand the load variation, and then, do the leveling by feeding some of the load either forward or backward. Under ESP, the basic load leveling method is to feed the load forward, since ESP aims to accomplish the zero missing products guarantee.

Four Principles:

1. Recognizing that productivity is chiefly determined at the production planning stage, it is therefore best optimized at the production planning stage.
2. Control of inventory (products, in-process goods, parts, materials) is best determined at the production planning stage.
3. There is no point in constructing an ESP that does not include improvement of production processes.
4. The Production Management Division must lead (by setting targets and goals and evaluating implementation) the improvement of production processes.

These four principles are indispensable for achieving and continuously maintaining the Six Guarantees. See Chapter 2.

Full work system: It is a term used in the Toyota Production System, and it is a method to prevent over production and to evaluate the balance of production capability among processes. In the full work system, standard volume of each process is established and if the volume is less than the standard, the production is continued, and if the volume is more than the standard, the production is stopped. In the Toyota Production System, over-production is considered the worst waste (muda).

Horizontal Keiretsu: Keiretsu is a group of Japanese firms that form long, continuous relationships. Some people consider such relationships as a source of the strength and competitive power of Japanese companies. There are different kinds of keiretsu. Some keiretsu are vertical, involving firms up and downstream from a firm that is usually a final assembler; some are horizontal, involving firms in different industries, but connected with Japanese main financial banks (also called a Zaibatsu, such as Mitsui, Mitsubishi, and Sumitomo). There are also other keiretsu which connect companies that are in the same industry (e.g. Hitachi and Matsushita).

Some criticize that keiretsu alienates overseas companies from entering the Japanese market. It is also true that some American automobile manufacturers were able to reduce cost and improve their quality by implementing keiretsu. In recent years, more and more Japanese companies are trying to work together with companies outside of their keiretsu to achieve world-class successful production.

Indirect division: In this book, indirect division means departments other than a manufacturing department. In general, indirect division includes production and distribution departments (production control, purchasing management, and distribution management), engineering departments (design, production engineering), research and development departments, marketing departments, and administrative management departments (finance, administration, accounting, etc.).

Inner circle (approach): When only a person, or people within a certain group, understands the contents, procedures, and the standards of the job. Sometimes, a person changes and revises the standard operation, and nobody else except this person understands how to do the job.

In-process inventory: It is also called "work in process." Items that are in a manufacturing process (from the beginning of the insertion process to the end of the final product process). It also includes works that are sitting in storage or between processes. In this book, 'in-process inventory" means the former: work in process.

Input sequence: Input sequence is the most efficient manufacturing sequence of products. The key to maximizing production efficiency is understanding the production conditions of the manufacturing department and creating a production plan that minimizes stoppages, speed, and defect loss in manufacturing processes. Chapter 1 and 2 in this book gives some concrete examples.

Insertion (method of scope): See Chapter 3, page 75 for details.
    Insertion method is one of the production planning simulation techniques of ESP to achieve the Six Guarantees.

Inventory: Inventory is stored items and resources necessary for production. There are four types of inventory: (1) purchased items such as raw materials and parts, (2) work in process, (3) final products for sale, and (4) spare items for maintenance, repair and tools.

JIT (Just in Time): A process aimed at increasing production efficiency and eliminating waste by producing the necessary items in the necessary volume at the necessary time. This idea was originated by Toyota Kichiro, the founder of Toyota Motors, and developed as a production system by his successors. Just in Time and Autonomation are the two pillars of the Toyota Production System.

JMAC (Japan Management Association Consultants): JMAC is a consulting company that became independent from JMA (Japan Management Association) in 1980. JMAC has 320 employees (1500 employees in the JMA Group) and its consulting services receive wide support and high reputation from its clients.

Kanban shortcuts: Kanban shortcuts means that using the Kanban just to order parts and materials without having any correct understanding of the Kanban system. Some examples of Kanban shortcuts are explained in Chapter 2. There are two reasons for Kanban shortcuts: (1) not recognizing that the Kanban system is one of the methods to actualize Just in Time, (2) not following the basic six rules of the Kanban system (1. Do not send defects to the downstream process, 2. Downstream process pick up items, 3. Only produce amounts that are picked up by the downstream process, 4. Average the production, 5. Kanban is a method for minor adjustment, 6. Stabilize and rationalize the process).

Keiretsu (structure): See "Horizontal Keiretsu", above.

Lateral deployment: To expand the improvements and knowledge achieved in a model to other product groups and other production processes. It is also called "horizontal deployment."

Limited inventory: This is also called "the controlled minimum inventory" or "the minimum required product inventory." Limited inventory consists of finished products.

ESP recognizes that suppliers cannot effectively control their production efficiently if they allow fluctuations in buyer order to be directly reflected in their production process. Limited inventory provides a buffer between buyer orders and the supplier's production schedule. Limited inventory enables the supplier to immediately respond to buyer orders or market needs while preventing missing items in deliveries.

Line stoppage (loss): There are two types of line stoppage loss: breakdown loss and changeover loss. Breakdown loss means that a machine is broken down and production is stopped. It also means that the machine's function is slowed down and it produces defects and decreases yield or speed of production. Changeover loss means how much time is lost from the time a line is stopped for changing the set-up of the product until the time the standard production is resumed for the next product.

Load leveling improvement: The Japanese translation of *Heijunka* (Production Smoothing, Level Production). It literally means to make flat, level and eliminate an imbalance. In manufacturing, production load leveling means to average both production volume and product variety in mixed production, and it is a technique used to adapt production to fluctuating customer's demand. It also means to create a production plan that enables the use of the production resources 4M (Man, Machine, Material, and Method) most effectively (Please see Chapter 2). In the Toyota Production System, load leveling (heijunka) is considered a foundation that supports two pillars— Just in Time and autonomation. For load leveling, at least two improvements are necessary: (1) improvement in production planning (establishing a production planning procedure that averaged/leveled both production volume and production variety); and (2) improvement in the manufacturing process (changeover improvements to achieve small lot production).

Management loss: Management loss means a loss generated when the production line is stopped due to the responsibility of the production management division (e.g. there are no operation standards, there are mistakes and omissions in operation, line stoppages due to missing parts and materials, or line stoppages due to meetings, etc).

Manufacturing process: This is a place in a manufacturing division where parts and products are manufactured. The processing and assembly stage in manufacturing systems where a series of manufacturing operations are conducted.

Materials resource planning (MRP): A basic planning method for determining materials requirements in a manufacturing environment. IBM introduced it in 1967 as a part of its information system called PICS (production information and control system) to control the manufacturing system.

Modularization: It is one of the techniques to correspond to the demand of various product specifications quickly, accurately and with low cost. There are two types of modularization: Simplifying parts and units, and then using combinations of simplified parts to provided a wide variety of product specifications; create advance functions using relatively simple lower level components for product configurations.

Multifunction design: One of the techniques to correspond the demand of various product specifications quickly, accurately and with low cost. Review products and production methods in terms of functions and structures and look for ways to achieve similar functions and structures using fewer parts or simpler manufacturing processes. See Chapter 4 for the four elements of multifunction design (Eliminate, Combine, Rearrange, and Simplify).

Multisource procurement: Ordering necessary parts to multiple suppliers or preceding processes of its company and have them deliver to a certain location.

Obsolete parts or materials: Parts and materials that have been stored for a long period of time and have become deteriorated and useless.

Optimization (of production): In this book, optimization means to actualize the six Guarantees in ESP, especially zero missing products, overproduction prevention, production load feed forward, production change adaptability and production efficiency maximization.

It also means to look at the whole manufacturing process and find out and improve the bottleneck processes (e.g. reducing the lead time, shortening changeover time for small lot production, etc.).

Order fluctuation: Order fluctuation happens in three situations: (1) order volume fluctuates when summarizing the orders from all customers periodically; (2) total order volume or order of the product item number from each customer is not stable; (3) there is a large difference between the content of the order instruction from the customer and the actual confirmed order.

Original concept one: This concept originated from the authors' consulting project in 1985. The basic idea is that a production geared strictly to customer needs is inefficient. This concept became the foundation of the four principles explained in Chapter 2. Please see Figure 1-1 and the explanation on page 3 in Chapter 1.

Original concept two: This concept also originated from the authors' consulting project in 1985. The basic idea is that to fulfill the production division's mission, daily production output and production sequences must be stabilized, with production output equalized among the various item numbers. Together with original concept one, this became the foundation of the four principles explained in Chapter 2. Please see Figure 1-1, and the explanation on page 3 in Chapter 1.

Overproduction prevention (planning): A planning procedure of ESP production planning, in order to prevent overproduction of products and parts. Under ESP, after a supplier performs the simulation under the first guarantee, it runs a second simulation. These results are then used to draft a new production schedule to prevent overproduction. If simulation results exceed the limited inventory, production is stopped to prevent overproduction.

PDCA: It stands for Plan, Do, Check and Action. It is also called PDCA cycle where managerial action is done by applying the sequence of plan, do, check, and action. It is a generalized version of Deming cycle that emphasizes the necessity of interaction among research, design, production and sales to improve quality.

Periodic fixed amount method: Ordering the fixed amount at a fixed period of time (e.g. newspaper delivery or milk delivery).

Periodic indefinite amount method: Ordering the indefinite amount (whatever is necessary at that time) at a fixed period of time. This method is used for ESP.

Planning delay loss: It means that the manufacturing division is able to carry out production, but it has to stop the production or set it on standby because the production schedule has not been issued. There are two examples of planning delay loss: (1) production adjustment loss: a loss generated by adjusting the production schedule depending on the supply and demand relationship, (2) Shut down loss: a loss generated by stopping the production or shutting down the factory due to a yearly inspection or construction.

Process capability: Production capability of the manufacturing process. An analysis of process capability means to investigate the capability of target processes to clarify any bottlenecks.

Post-improvement conditions: A method to determine the volume of limited inventory. It means that a supplier should expect some improvements in their processes and set the target of limited inventory a certain percentage less than the current inventory levels.

Principle one: One of the Four Principles that is behind the ESP. Four principles are absolutely necessary in order to achieve the Six Guarantees. Principle one is to recognize that productivity is chiefly determined at the production planning stage; it is therefore best optimized at the production planning stage. See Chapter 2.

Production planning proposals (system): The ESP production planning drafting system that uses the computer information system. By using the production planning proposal system, a company drafts an optimum production plan and gives instructions to each department accordingly. The results are gathered as actual information and reflected on the next production plan.

Production sequences: Sequences to make products as indicated in the ESP production pattern. Production sequences are based on the principle of maximizing the efficiency of their own manufacturing processes. The supplier creates ESP production patterns that include production sequences based on the principle of optimizing its own manufacturing processes. See also "input sequences."

Range extension: It is one of the techniques to correspond to a demand of various product specifications quickly, accurately and with low cost. It means to extend the range of performance features covered by individual parts in order to lower costs for changeover, equipment, dies, jigs, tools, etc.

Short lead-time bottlenecks: A bottleneck process (also called a "critical process") is a process where it is not possible to shorten the lead-time any more.

Supply chain management (SCM): A management method using the IT to control orders, materials, parts, inventory, production, and product delivery between customers and its own company. It is also said that it was born from a research of Japanese style management done by an American manufacturing industry.

Semi-direct operations: Operations that support, manage, and monitor direct operations (e.g. manufacturing operation, processing and assembly operation, etc). In this book, preparation and transportation of parts and materials, monitoring and supervising manufacturing process are described as semi-direct operations.

Six Guarantees: 1) Zero missing products guarantee, 2) overproduction prevention guarantee, 3) logical production management guarantee, 4) production efficiency maximization guarantee, 5) production change adaptability guarantee, and 6) production load feed forward guarantee. These Six Guarantees are the achievement goals of ESP. By achieving the Six Guarantees, companies can increase a profit and receive customer's trust. See Chapter 3 for details.

Stops: One of the production planning simulation techniques to achieve the Six Guarantees in ESP. There are two purposes in stops. (1) The supplier will perform a stop if, after incorporating the buyer's order information into the relevant ESP production pattern and running a simulation, the simulation indicates you will be overproducing certain item numbers. (2) The supplier will perform a stop if the simulation, based on the ESP production pattern and calculated product inventory (estimated inventory), selects additional item numbers that need to be produced in response to a buyer's rush orders.

Store inventory: The term used in the Toyota Production System. It is an inventory of parts that are already processed in the processing line. Store is an area where the inventory is placed, and people in the downstream process can come and pick up necessary parts.

Supplier: A supplier is a company that supplies parts and materials to a final company (a company that manufactures final products for consumers).

Supplier's rationale: Unless a supplier is completely dedicated to only one buyer (to be more precise, unless a supplier produces parts only for the one particular line of the buyer), it is impossible for the supplier to follow the buyer's order, meet the buyer's need and to achieve high production efficiency. The buyer makes demands to the supplier as if buying parts from a vending machine. The supplier cannot respond to such demands.

Synchronization: In general, as represented in Just in Time, synchronization means to acquire and supply the necessary products in the necessary volume at the necessary time. Synchronization in ESP has the same meaning. Please see Chapter 3 for the information on synchronized and equalized production systems.

Systemization: Systemization in this book (Chapter 3) is one of the primary functions of production control. Primary functions of production control are divided into three parts: planning, instruction, and systemization. The general meaning of systemization is to unify things that have a tendency to be separated or scattered. Therefore, systemization of production control is to follow-up and manage the progress of the production planning, gathering information and results, and shop floor management.

Toyota Production System: A production system that was created by Toyota Kiichiro, a founder of Toyota Motors. It is a very practical and concrete production method. TPS has two pillars: Just in time and autonomation. Those pillars are supported by the production smoothing (heijunka) as the foundation.

Unusual flow items: Product item numbers that are seldom in production. Although an input sequence and production size are set for unusual flow items, unusual flow items do not have their own production schedule in the ESP production pattern.

Usual flow items: Product item numbers that are frequently in production. Any item registered as an item that is usually in production as part of the ESP production pattern is a usual flow item.

Velocity loss: There are two types of velocity loss: minor stoppage loss and speed loss. Minor stoppage is when a machine is stopped for a short period of time due to a temporary problem instead of a breakdown, or a machine is in an idle state (a machine is operating, but not processing). Speed loss is generated when a machine is operated in a slower speed than a designed speed, because if a machine is operated in a designed speed, it creates mechanical and quality problems.

Vertical keiretsu: See Horizontal keiretsu or Keiretsu, above.

Visible inventory: To make the product item numbers and the amount of the inventory visible in order to manage the inventory correctly.

Zero inventory: Zero inventory is one of the ultimate goals for all kinds of production systems. Basically, having inventory is not good, and it is best not have any. Many companies are trying to move towards "production to order" systems by applying IT and speed up the information flow. ESP's ultimate goal is also zero inventory. As stated on page 62 in Chapter 2, it is very difficult to achieve zero inventory with the Kanban system.

Index

About the Authors

Toshiki Naruse is a principal consultant at Japan Management Association Consulting (JMAC), with 40 years of experience in manufacturing. Naruse originated the concepts that developed into the Equalized and Synchronized Production system (ESP). Prior to joining JMAC, he worked for 14 years in a leading automobile parts supplier company in Japan, in positions including production control and machine design. Naruse has researched and practiced consultation in Total Productive Maintenance (TPM) since 1983 and in ESP since 1985. He has successfully applied ESP through his work with numerous automobile parts manufacturers, and he has expanded the application in other industries including general-purpose machinery and parts, precision parts, food products, and medical equipment.

Kenichi Morii has over 20 years of experience consulting in production system improvement. He has provided expertise in implementing ESP for over 30 companies in both discrete and process manufacturing. He is a senior consultant with JMAC.

Kunio Shibata's manufacturing career began in production control at a manufacturer of time management and environmental management equipment. He has extensive experience in delivery, production, and purchasing operations. In particular, he has practical experience with production control functions and operations for standard products and job-order production. He joined JMAC in 1986 and is currently a chief consultant.

Tsutomu Iwabuchi has extensive experience in overall equipment development and design for autonomation and energy conservation. Prior to joining JMAC in 1983, he worked in equipment development at a manufacturer of electrical appliances. At JMAC, Iwabuchi has been involved in building production systems for job shops, improving efficiency in engineering, and conducting research in CAM and advanced production technology. He is currently a chief consultant with JMAC.